The Dynamics
of Progress

The Dynamics

TIME

of Progre

METHOD, AND MEASURE

Samuel L. Macey

THE UNIVERSITY OF GEORGIA PRESS   ATHENS & LONDON

Paperback edition, 2010
© 1989 by the University of Georgia Press
Athens, Georgia 30602
www.ugapress.org
All rights reserved
Designed by Louise M. Jones
Set in 10/13½ Linotype Walbaum
Printed digitally in the United States of America

The Library of Congress has cataloged the hardcover edition of this book as follows:
Library of Congress Cataloging-in-Publication Data
LCCN Permalink: http://lccn.loc.gov/89031886

Macey, Samuel L.
The dynamics of progress : time, method, and measure / Samuel L. Macey.
xiv, 273 p. : ill. ; 23 cm.
ISBN 0-8203-1159-6 (alk. paper)
Includes index.
Bibliography: p. [243]–261.
1. Technology and civilization. 2. Progress. 3. Rationalism. 4. Time measurements—History. I. Title.
CB478 .M244   1989
304.2'3 20   89031886

Paperback ISBN-13: 978-0-8203-3796-8
ISBN-10: 0-8203-3796-X

British Library Cataloging-in-Publication Data available

The illustrations on the half title and title pages are from Diderot's Encyclopedia, volume 4; used by permission of the Hargrett Rare Book and Manuscript Library, University of Georgia Libraries.

For Elizabeth and Caroline

Contents

	List of Illustrations	ix
	Preface	xi

PART I   RATIONALIZING TIME

I	Time and Clocks: The Divisions of the Day	3
II	Calendars: The Days of the Year	25
III	Chronology: The Years of the World	41

PART II   RATIONALIZING MEASURES AND COMMUNICATION

IV	The Ascendancy of the Metric System	65
V	The Ascendancy of Hindu-Arabic Numerals	82
VI	The Ascendancy of English	112

PART III   RATIONALIZING PRODUCTION

VII	Great Britain and the Industrial Revolution	139
VIII	North America and the World	155
IX	Retail Distribution and Finance	181

PART IV RATIONALIZING HUMAN BEINGS

X Labor and the Individual 201

XI Human Equality 218

Conclusion 231

Notes 243

Index 263

Illustrations

Figure 1. Copy of a mid-eighteenth-century painting by François Boucher — 16

Figure 2. William Hogarth, *An Election Entertainment*, 1753–54 — 33

Figure 3. British parody of Jakob Köbel's *Geometrei* — 70

Figure 4. Family Tree of the Indian Numerals — 89

Figure 5. Chinese abacus — 90

Figure 6. Swift's dedication to an early version of "Verses on the Death of Dr. Swift," 1733 — 120

Figure 7. Illustration from *Das Feuerwerkpuch*, ca. 1450 — 141

Figure 8. Girls' dining room of the H. J. Heinz Company, 1914 — 204

Figure 9. Gustave Doré, *Wentworth Street, Whitechapel* — 209

Preface

This study is concerned with the very widespread human desire for material progress, and the history of our endeavors to achieve such progress in the period since England's Restoration of 1660. *The Dynamics of Progress: Time, Method, and Measure* is the third in a series of books in which I have been concerned with the question of time. *Clocks and the Cosmos: Time in Western Life and Thought* (1980) deals with the extensive influence of clocks on the language and imagery used by philosophers, theologians, and poets during the period 1660–1860. *Patriarchs of Time: Dualism in Saturn-Cronus, Father Time, the Watchmaker God, and Father Christmas* (1987) is concerned with the way in which our attitudes toward the personifications of time have changed during the past four thousand years. Both of the previous books were involved with the rationalization that has taken place in our measurement of time since the invention of mechanical clocks some seven hundred years ago. This book uses temporal rationalization as the starting point for examining Western technology and the human world that is affected by technology.

It is now difficult for us to comprehend that in the time of Chaucer, who died in 1400, the length of the hour varied according to the latitude, the time of the year, and the day or the night. This was because the so-called temporal or variable hours were one-twelfth of the day and one-twelfth of the night. In London, at a latitude of approximately fifty-one and a half degrees north, the daylight varies from sixteen and a half hours at midsummer to seven hours and forty-five minutes at midwinter. The figures translate into eighty-two minutes and thirty seconds of clock time per temporal hour at midsummer, and thirty-eight minutes and forty-five seconds at midwinter. Such an irrational division of the day might do well enough in an essentially agricultural civilization. But it would be quite inappropriate for our present urban society, with its quartz watches and its domination by time. Since Chaucer's day there have been several rationalizations in the division of time. By 1967 the calibration of the second itself was divorced from the rotation of our earth, which is no longer considered an accurate enough measure.

It is generally recognized that modern science bases its predictions on accurate measurement and lucid communication. My awareness of the remarkable rationalization that has taken place in the measurement of time made me wonder whether other forms of measurement and communication might not have been similarly rationalized. The results, as reported in this book, are very enlightening. They show an almost symbiotic relationship between the rationalization of measurement and communication, and related developments in pure and applied science, as well as in technology.

Part 1 of *The Dynamics of Progress* deals with the question of rationalizing time. I feel that this is the key to understanding modern Western society. So many of our new disciplines, since the Restoration of 1660, are concerned with progress through time. These include dynamics, calculus, paleontology, archaeology, geology, anthropology, Freudian psychology, Hegelian and Marxian dialectics, Darwinian biology, Einsteinian physics, and many more disciplines that represent a breaking out from the earlier static and feudal society. The rationalization of time measurement is dealt with in three chapters, each concerned with a major measure of time: the day, the year, and the period since the creation. Part 2 deals with three further major areas in which measures and communication have been rationalized. The first chapter is concerned with measures of length, area, weight, and volume. It shows that the rationalization has here been just as radical as that which occurred in the measurement of time. The second and third chapters in Part 2 deal respectively with the rationalization of numbers and of language.

Part 3, also in three chapters, is concerned with the way that the rationalization thus far considered acts as the underpinning for the modern rationalization of production, which we call the industrial revolution. During the past three centuries, the greatly expanded production of material goods in the Western world has, however, not been an end in itself. Part 4 deals with the rationalization of human beings, both as producers and as consumers of today's exponentially increasing quantities of goods and services.

Human beings have, understandably, reacted against all forms of rationalization, including their "Romantic" reaction against *ratio*, or reason, itself. I have indicated some of those reactions throughout the book. My main purpose, however, has been to show the remarkable extent of the rationalization that has occurred during the past three hundred years, in a wide range of human endeavor. Clearly there must be a paradox when people want more and more material goods and yet complain bitterly about the inevitable rationalization of their work and their lives, as well as the increasing contamination of their earth. As we approach the era of the global village, this is a subject with which we are all very much concerned, but about which we have a whole spectrum of conflicting views. As far as possible, though not always with success, I have tried to play down my own opinions and concentrate on the subject of rationalization. I am aware that rationalization is a potentially raging theoretical issue that will play an increasing role in our sense of ethical values. But that is not my direct concern in this study. My hope is that by concentrating on the historical process I will remain as free from polemic as possible, while providing new insights into the breadth and depth of the rationalization that has already occurred. In dealing with rationalization, I have also tried to remember that what we now consider irrational may have been thought rational (or an act of *ratio*) in the past. Since our sense of rationality may have changed continuously, I have, throughout the study, attempted to judge what is rational or irrational by modern standards.

The title of this book suggests that the dynamics of progress are dependent upon measure, method, and time. Without increasingly accurate measure, today's science could never have been developed. Method and time are equally essential for modern progress. Time and method studies—by their many names, as we shall see—are a key to the rationalization that has given us Western production methods. The root meaning of rationalization is the application of *ratio*, or reason, to simplify and standardize human activities over time. Time and method should therefore remind us how carefully we must now handle the key tool of rationalization through which the fruits of Western technology have thus far been produced. This study should help us in making the difficult decisions that are crucial if humanity and its offspring are to continue to inherit the earth.

For both the content and the form of this book, I am obliged to many people over many years. My interest in time goes back to five years at sea in the Royal Navy during World War II and a business career thereafter that included the wholesale distribution of clocks and watches. I am indebted to the University of Victoria and the Canada Council for financial aid related to the study of time; to the Institute of Management Services, of which I have been a member and a fellow for thirty years; and to Corpus Christi College, Cambridge, which facilitated my early research into matters temporal by electing me a visiting fellow in 1972. I have acknowledged in the notes ideas that have appeared in my earlier books and articles on literature, productivity, and time, but I would like to mention in particular this book's obligations to "Literary Images of Progress: The Rise and Fall of a Western Ideal," in *Study of Time V* (Amherst: University of Massachusetts Press, 1986) and "Work Study Before Taylor," *Work Study and Management Services* 18 (October 1974). I am indebted to three research assistants—Susan Cripps, Wendy Bond, and Jane Sellwood—who, over several years, have given invaluable help with this project. I am especially grateful for the encouragement and assistance of the members of the editorial staff at the University of Georgia Press. I have benefited greatly from the discussions and writings of many members of the International Society for the Study of Time. These include George Ford, Adam Mendilow, David Park, Lew Rowell, and the late Nathaniel Lawrence, as well as J. T. Fraser, who has been a mentor to all of us. I have also received much help from the British Horological Institute, the librarians of the McPherson Library, George Marsden, Fenwick Lansdowne, and my colleagues at the University of Victoria. Finally, and above all, I am grateful to my wife. Since this is a book that shows how the past may influence the future, we are dedicating it to our two daughters in the hope that they will leave to their children a saner and safer world than we are leaving to them.

Rationalizing Time

I Time and Clocks
The Divisions of the Day

Our lives are now so precisely regulated by time measurement that we have become almost oblivious to the fact that existence was not always like this in the Western world. In more primitive countries, older methods of time reckoning have carried on into our own century. Writing in 1920, Martin P. Nilsson tells us that "in Madagascar 'rice-cooking' often means half an hour, 'the frying of a locust' a moment." He tells us further that the following expressions for a period of time were being used by the Malays, the Javanese, and the Achanese: "A blink of the eyes (literally), the time required for chewing a quid of *sirih* (about five minutes), the time required for cooking a *kay* of rice (about half an hour), for cooking a *gantang* of rice (about an hour and a half), half a day, a 'sun dark', i.e. a complete day and night."[1] In the world that is fast becoming a global village, such localized forms of time measurement have already, or will very soon have, disappeared. In more advanced civilizations, the attempt to divide the day rationally has a much longer history. Generally, the day was divided into twelve hours, and so was the night. Since the length of the day or night varied considerably, however, between winter and summer, the length of such unequal or "temporal" hours varied accordingly. At Cairo, in the latitude of about thirty degrees north, there are approximately fourteen hours of daylight by the clock at midsummer, and ten hours at midwinter. This meant that at midsummer the hour contained seventy clock minutes during the day and fifty minutes during the same night. But at midwinter in that town the hours contained fifty clock minutes by day and seventy minutes by night. The irrational division of time was even greater in London, at a latitude of approximately fifty-one and a half degrees north. There, Chaucer was still explaining the method for determining temporal hours in his *Treatise on the Astrolabe* (ca. 1391). In London, the daylight varies from sixteen and a half hours at midsummer to seven hours and forty-five minutes at midwinter. Thus the length of a temporal hour in London varied between about eighty-two and one-half and thirty-eight and three-quarters minutes of clock time.

Though he notes that the "sense of the 'hour' existed earlier," E. J. Bickerman relates that "as early as *c*. 2100 BC, the Egyptian priests were using the system of twenty-four hours: ten daylight hours, two twilight hours, and twelve night hours. This arrangement . . . gave way *c*. 1300 BC to a simpler system which allotted 12 hours to the day and 12 hours to the night." He adds that the Babylonians similarly divided the day and the night by twelve, and that the Greeks, according to Herodotus, learned this arrangement from the Babylonians.[2] Such methods of dividing the day were, of course, quite arbitrary. The Chinese, like the Japanese, divided our twenty-four-hour day into twelve subunits, and the Hindus came to use sixty subunits.[3]

The Japanese use of temporal hours is particularly interesting. Though time reckoning by equal hours was introduced in Europe shortly after the advent of mechanical clocks, in Japan temporal hours continued in use until as late as 1873. This was nineteen years after Admiral Perry concluded the first treaty between the United States and Japan, on March 31, 1854, in Edo Bay, the modern Tokyo. That action marked the opening of Japan to Western influences. According to F. A. B. Ward, the Japanese appear to have included a considerable period of morning and evening twilight in their measurement of the day. By this reckoning, at their latitude of approximately thirty-five degrees north, "day" and "night" were of equal length at midwinter, whereas at midsummer the day was some two and a third times as long as the night.[4]

Though temporal hours may very well seem irrational to us, they were of far less inconvenience when more than 95 percent of the people lived on and from the land. With almost no artificial light, virtually all of the productive work was performed during daylight hours. It therefore appeared natural enough to use the sun's shadow to mark off the passage of the daylight. In such a society, the daylight hours are bound to predominate. Hence the awkwardness in our language when it came to differentiating between the "day" of daylight and the day of twenty-four hours; hence the reckoning of the day from sunrise or sunset, as still reflected in the Jewish theological day, which begins with the sunset of the previous evening; and hence the persistent naming of our Western Deity with the attribute of day. In the Greek pantheon, Uranus (Greek *ouranos*, "the sky," the air god of a nomadic people) married Gaia ("earth," the earth goddess of an agricultural people), and produced Cronus (who by the sixth century B.C. had taken over the attributes of a god of time).[5] Zeus—the son of Cronus who castrated and succeeded his father, as Cronus had earlier done with Uranus—was the god of day (Zeus, *Deus*, deity, *Dieu*, *dies*, day).

Though the cycle of the whole day as a unit of time was self-evident, the number of units into which it might be divided was quite arbitrary. We have already noted that the Hindus used sixty units and the Chinese and Japanese used twelve units for our twenty-four-hour day. In addition, the Saxon sundials divided the total day into four "tides," a practice still reflected in our terms *noontide*, and *eventide*. The requirements of war long ago led to the division of day and night into watches. According to Bickerman, "The Babylonians, the Old Testament and Homer . . . had three watches during the day and three more during the night, while the Greeks and the Romans later adopted the Egyptian system of four watches . . . which was also widely used in civil life to indicate parts of the night."[6]

Five years of watchkeeping at sea during World War II have demonstrated to the author how older systems continue in modified form long after their origins have been forgotten. The Royal Navy's *Manual of Seamanship, 1937* lists the following, under the heading "Time":

The day of 24 hours is divided into seven watches: the three different methods of describing them are:—
 0000 to 0400   Midnight to 4 a.m.—middle watch
 0400 " 0800   4 a.m. to 8 a.m.—morning watch
 0800 " 1200   8 a.m. to noon—forenoon watch
 1200 " 1600   Noon to 4 p.m.—afternoon watch
 1600 " 1800   4 to 6 p.m.—first dog watch
 1800 " 2000   6 to 8 p.m.—last dog watch
 2000 " 0000   8 to midnight—first watch[7]

This table reflects many of the changes in the division of time with which we are concerned. Though it deals with equal hours and uses the unambiguous twenty-four-hour day as employed on the continent of Europe, the Royal Navy's system has adapted the old six-watch division of four hours to the watch, as used by the Babylonians and found in the Old Testament and in Homer. The division of the earlier 1600-to-2000 watch into a first and second dog watch partitioned the day into seven sections. This modification was clearly of benefit to the needs of men who must serve "watch and watch about" during the time that they are at sea. The listing of the watches—beginning with the middle watch from midnight to 4:00 A.M., and ending with the first watch from 8:00 P.M. to midnight—is also revealing. The placing of the "first watch" at the end of the list must surely suggest its derivation from an older system in which the day began at sunset rather than at midnight.

EARLIER TIME-MEASURING DEVICES

As we have previously noted, "sundial time," in one form or another, long controlled time measurement. But the use of sundials had progressed considerably from earlier days when trees, columns, and even obelisks had been employed. In lands where the sun was virtually overhead, men measured the division of the daytime by the length of the shadow; in more northerly latitudes they used the direction of the shadow. Though the earliest extant shadow clock is a fragment from Egypt of ca. 1500 B.C., the sundial does not seem to have appeared in Rome until about 290 B.C. Vitruvius's *De architectura* (first century A.D.) indicates, however, that both the types and the numbers of sundials in Rome had, by his time, proliferated considerably.

There were, of course, earlier methods that might more naturally have denoted equal time had they not been subservient to the domination of "sundial time." The Cairo Museum has an early Egyptian water clock of ca. 1400 B.C., of which there is a cast in the Science Museum, London. The clock measures time by slowly losing water from a hole near the bottom. For each of twelve months, there is marked on the inside a separate series of water levels that divides the average night for that month into twelve equal parts. This suggests that the need to provide hours of different lengths for day and night must have considerably complicated a relatively simple method for denoting time. The same would have been true of the later complex clepsydrae (or water clocks), which involved wheelwork, jacks, and other automata. These provided a transition to the public mechanical clocks of the fourteenth century.

Short periods of time must have been measured equally, even before the advent of equal hours. Primitive measurement—such as the time taken to walk to a particular place or to cook a known quantity of rice—is not readily adaptable to temporal hours, nor are King Alfred's candles, the sinking-bowl type of water clock, or the water clocks known to have been used for limiting lawyers' speeches in classical times. Both mechanical clocks and sandglasses had been developed in the early fourteenth century.[8] Since all are called *horloges*, we cannot readily differentiate literary references to water clocks and sandglasses from those to mechanical clocks, except through the context. As a result, we are surprisingly uncertain about the precise beginnings of both sandglasses and mechanical clocks. I have no knowledge of early sandglasses being adapted to temporal hours, but for some time it was necessary to accept the inconvenience of adapting clocks to indicate temporal or variable hours.

From their beginnings and for some three hundred years, the main escapement that controlled mechanical clocks was the verge and foliot. That escapement was not isochronous in any modern sense, but nonetheless it was an invention of great importance. In the verge-and-foliot escapement, the foliot bar oscillates back and forth, moving with it the two pallets on the verge. The verge and foliot are actuated by the crown wheel and yet regulate the escape of the crown wheel through the two pallets. Though the crown wheel was first driven by weights, the invention of the spring drive, around the middle of the fifteenth century, made watches, or portable clocks, feasible for the first time. The speed at which the foliot bar moves in the verge-and-foliot escapement can be adjusted by changing the positions of the two small weights that hang at each end. If the weights are moved toward the center, the clock gains.

Early Japanese clocks demonstrate the complicated system necessary for adjusting the verge-and-foliot escapement to measure temporal hours. From 1600 on, when Dutch navigators were in regular contact with the Japanese, the mechanical timekeepers taken there as gifts or for trade seem to have been lantern clocks with a verge-and-foliot escapement. Thanks to the pressure from mechanical clocks, sandglasses, seamen, and urbanization, England had, by the time of Shakespeare, for the most part converted to equal hours. Bickerman notes, however, that the temporal or variable hour "persisted in some parts of the Mediterranean world well into the nineteenth century."[9] Since the Japanese rejected Western influence after importing verge-and-foliot clocks, they retained temporal or variable hours until the latter part of the nineteenth century. Thus their situation represents a valuable time warp for horological historians. They were obliged to build their own clocks with twin escapements, one for the day and one for the night. The changeover from day or night hours was automatic and effected by the striking mechanism. As Ward puts it, "The only adjustments then necessary were to move the weights of each foliot as the seasons changed; this operation would be carried out, probably by a clockmaker, about once a fortnight."[10]

Several years ago I was taken to the Daimyō Clock Museum, in the back streets of Tokyo. After removing my shoes, since this was seemingly holy ground, I was allowed to inspect a remarkable collection of such clocks. For those who do not wish to travel this far, there is a similar Japanese "lantern" clock in the Science Museum in London. When European time reckoning was introduced to Japan in 1873, modern clocks, which did not require the services of a clockmaker every two weeks, began to be made. Indeed, the first clock factory was set up in 1875.

EARLIER DIVISIONS OF THE DAY

We now take the division of the day into twenty-four equal hours so much for granted that it is easy to forget how many and how varied were the divisions of the day before this was accomplished. Even the word *hour* did not originally mean the twenty-fourth part of a day. The Greek *ora*, from which we derive our *hour* via the Latin *hora*, originally referred to a season, and then to the fitting or appointed time. In Greece, according to Bickerman, "The sense of 'hour' is first attested in the second half of the fourth century BC."[11] The primacy of the hour in time measurement is demonstrated by the root meaning of such temporally related words as *horologium, horloge,* and *horology*.

It has been argued by Lewis Mumford, among others, that orderly, punctual life in the West first took shape in the monasteries. Eviatar Zerubavel has demonstrated that the daily order of work and prayer introduced by Saint Benedict, in the sixth century, still had to be extremely flexible by our standards. In part, this was probably because human beings were then less amenable to time scheduling. In addition, the hours with which they were dealing were still variable throughout the course of the year.[12] The eight canonical hours are matins, lauds, prime, terce, sext, none, vespers, and compline. Matins was originally celebrated at midnight, or shortly thereafter, and compline was celebrated last thing before going to bed. Prime, terce, sext, and none once represented, as their names suggest, the nine daylight hours. These were, at first, 6:00 A.M., 9:00 A.M., noon, and 3:00 P.M.

The tidy modern mind might be excused for thinking that the canonical hours of the church were a system of eight three-hour watches. However, they have been much modified since their beginnings—and even as recently as 1911, 1955, and 1960—in order to accommodate them to the more rationalized temporal measurements of modern life. The irony is that though the *horarium* or monastic daily schedule (with its related *Rule of Saint Benedict*) played an important part in regularizing life, the related canonical hours have been continually modified by the needs associated with secular time measurement.

In the late fourteenth century the English were still using temporal or variable hours, but people were beginning to become more time conscious. Chaucer does not mention weight-driven clocks in his canon, though the earliest extant examples in England date from the latter part of his lifetime. The Salisbury Cathedral clock is thought to date from 1386, and the Wells clock from 1392. In his *Treatise on the Astrolabe* (ca. 1391), Chaucer feels the need to explain the difference between "unequal" (or temporal) hours and the "equal" hours used in astronomy. John Trevisa, in his translation *On the Properties of Things* (1398), lists the average number of equal hours in the night and day for each month. Under January, for example, he says that "this moneth hath longe nightis of sixtene houres and shorte dayes of eighte houres," whereas December's "nyght hath xviii houres and his day hath sixe." Trevisa is aware of the awkwardness in differentiating between the daylight day and the twenty-four-hour day: "Somme dayes is artificial and somme naturel. Artificial day is the space in the whiche the sonne passith aboute in oure sight fro the est to the west. . . . A naturel day is the space in the whice the sonne passith aboute out of the est by the west into the est agen; and suche a day conteyneth xxiiii houres."[13]

Trevisa clearly specifies "a naturel day" when he is listing the temporal divisions: "a moneth conteyneth foure wekis, and a woke [sic] seuene naturel dayes, and a day foure quadrantis, and a quadrant sexe houres, and an hour foure poyntis, and a point ten momentis, and a moment twelue vncis, and an vnce xlvii. attomos; and attomos is no ferther departid for his schortnesse."[14] This quotation, which seems to be related to the table of Papias in Du Cange, contains a number of elements with which we will be further concerned. Though Trevisa's use of Middle English characters has been silently eliminated, one can still readily observe that the language he employs has been considerably rationalized during the past six hundred years.

If we examine Trevisa's divisions of the hour, we may be surprised to learn that the quarter of an hour is denoted as a point in time. From the term *point*—which we would consider a very long period of time for such a designation—we presumably derive such expressions as punctuality, punctiliousness, "the appointed time," and "at the point of death." The moment (or one and a half modern minutes) derives from the term *movement*, which, like the atom, relates time to a short physical motion or movement. The ounce, or twelfth part of a moment, we shall meet again when dealing with the rationalization of weights. The atom, which the *Oxford English Dictionary* derives from the Greek for "the twinkling of an eye" or "indivisible," will remind us that time and space were interrelated long before our own century. Clearly, given the inaccuracy of early clocks, moments, ounces, and atoms of time were little more than theoretical concepts. However, the verge-and-foliot escapement is the key to the rationalization of variable, or temporal, into invariable or equal hours. As late as 1516, Thomas More still feels the need to stress that his Utopians employ equal hours, but by the time of Shakespeare the use of equal hours is assumed.

ACCURATE CLOCKS AND NAVIGATION

The verge-and-foliot escapement may have been applied first to regulate the work and prayer of monks. But very soon it was regulating all aspects of human life and eventually forced a rationalization of the variable or temporal hours themselves. The next great step in the rationalization of time derived from a very different source. We owe the pendulum escapement to the persistent demands of astronomers. This demand occurred after the heliocentric views of Copernicus, published in *De revolutionibus orbium coelestium* (1543), brought about a revolution in astronomy. As a result, astronomers became increasingly insistent that clockmakers should provide them with much greater accuracy in time measurement. By the end of the sixteenth century, Jost Burgi was trying to meet the standards required by Tycho Brahe and Johannes Kepler. His crystal clock (ca. 1600) provided separate rock-crystal dials for hours, minutes, and seconds; this is one of the first recorded uses of the second hand.

Galileo was also much concerned with the accurate measurement of time. In 1581 or 1582 he is said to have noted the isochronous quality of a swinging lamp in the cathedral at Pisa, and to have tested this against his pulse. Though pendulums maintained by hand were soon being used by astronomers for their observations, Galileo does not appear to have related his idea to clockwork for many years. In this sense the clockwork of a pendulum clock may be seen as the automaton that takes the place of the astronomer. Similarly, the earliest mechanical clocks may be considered automated versions of the keeper of the bell or *Glocke*, which gave us the term *clock* in the first place. (The automaton jacks that ring the bells on early clocks illustrate this concept.)

Although Galileo, just before his death in 1642, seems to have conveyed his ideas on the pendulum escapement to his son Vincenzo, those ideas were not brought to fruition before his son's death. Instead, Christiaan Huygens must be credited with the pendulum clock, which is one of the world's most important inventions. In 1657 he succeeded in using the clock as a mechanical method for both maintaining the movement of the pendulum and counting the number of its relatively isochronous swings. As he wrote in his *Horologium* (1658), "Astronomers, certainly, are adopting it [the pendulum clock], so that henceforth there will be no troublesome urging of pendulums nor watchful counting required."[15]

Before 1657 clocks did not generally keep time more closely than about ten or fifteen minutes per day, but now their accuracy had been increased more than sixty-fold by a single invention. For the first time, clocks were sufficiently accurate for the needs of urban man, and this resulted in some industrial espionage that was remarkable even by modern standards. Huygens assigned his rights in the pendulum clock to the tradesman Salomon Coster, who took out a patent in Holland on June 16, 1657. In September 1657 John, the eldest son of the clockmaker Ahasuerus Fromanteel, left England to work with Salomon Coster. The first known English pendulum clock is signed on the backplate, "A. Fromanteel, London fecit 1658."[16] This opens the century of what I have called the British horological revolution. It begins with clockmakers satisfying the needs of astronomers, but its greatest pressures for further innovation came from the requirements of navigators. In 1761, at the end of the British horological revolution, John Harrison's fourth chronometer erred by no more than fifteen seconds after a five-month journey to the West Indies and back.

As David W. Waters has argued, "It is time which makes modern civilization practicable. But it is the provision of accurate time in ships at sea which lies at the core of civilization."[17] Eric Bruton makes an even stronger case for the importance of the twenty thousand pounds that the British Admiralty offered for the discovery of a method to ascertain the longitude at sea: "The act of 1714 caused the same kind of surge of scientific effort that space research does today, and was in many ways responsible for the Industrial Revolution that followed. The invention of the marine chronometer, for which it was directly responsible, resulted eventually in the domination of the world by the British Fleet, the expansion of trading and the acquisition of the British Empire."[18] I have myself written elsewhere of the importance of the marine chronometer in the hands of such British seamen as Cook, Bligh, and Vancouver.[19]

The need for an accurate marine clock had been foreshadowed more than two centuries earlier by Gemma Frisius, the teacher of Mercator. But before Huygens's pendulum clock—and his subsequent spring-balance escapement of ca. 1674, which permitted marine clocks to be portable—such a timepiece would not have been possible. The reason why an accurate marine clock is needed to discover the longitude but not the latitude is as follows: when the sun is directly to the north or the south (local noon time), a seaman can ascertain his latitude by measuring the angle between the sun, the ship, and the horizon. Thus with the aid of a sextant and a compass he knows his latitude; in other words, he knows how many degrees he is north or south of the equator. Because, however, the earth is spinning on its axis, it is an incomparably more difficult task for the seaman to compute his longitude, his position east or west of the Greenwich meridian. Yet the earth spins at a relatively even rate through 360 degrees, during a period that we have divided into twenty-four hours. Therefore, if a seaman had an accurate clock regulated to the time at Greenwich, England, he could discover his longitude. For example, if he found that (in the Atlantic) his clock indicated 2:00 P.M., when the sun was directly to the south (local noon time), he would know that he was somewhere on the longitude 30 degrees (360 degrees divided by the difference in time) west of Greenwich. Since, with a compass and sextant, the seaman could also ascertain the latitude, he would thereby know his precise position.

There were at least two other viable methods for ascertaining the longitude—the "lunars" introduced by Nevil Maskelyne (1732–1811), and "the clock in the sky," based on the eclipses of Jupiter's moons, introduced by Giovanni Domenico Cassini (1625–1712) as director of the Paris Observatory—but neither system could compete with the chronometer for ascertaining the longitude at sea.

From the fourteenth century onward, seamen in the Mediterranean were finding their way by dead reckoning. For this, charts provided the bearings or "winds," the compass gave the direction, the sandglass furnished a timekeeper that was unaffected by the motion of the sea, and the ship's log supplied an approximate indication of speed. An early form of log was a "chip log" attached to a knotted cord. Thus, in England, one was later able to measure the speed of a ship in knots by counting the number of "knots" that ran out in the time that it took to empty a twenty-eight-second sandglass.

The fall of Constantinople to the Turks in 1453 changed drastically the trading routes to the East. As a direct result Vasco da Gama reached Calicut, in India, on May 20, 1498, and Christopher Columbus set out, on August 3, 1492, to reach the Far East by sailing westward. Thus began the era of long ocean voyages from Europe, for which the discovery of the longitude would become increasingly important. The wealth that the Far East and the Americas provided—in such commodities as spices, silver, and gold—is now part of the history and the myth of the Western world. It was a full century before the British began to get their share, and more, of that wealth, first as privateers, then as traders, and, finally, as settlers. The voyage literature of Samuel Purchas and the novels of Daniel Defoe still give us a very thrilling insight into that period. Waters describes the remarkable change in England's position as a seafaring nation: "as late as 1568 probably only one English seaman was capable of navigating to the West Indies without the aid of Portuguese, French, or Spanish pilots. Yet by the time of the Armada [1588] . . . Englishmen had gained a 'reputation of being above all Western nations, expert and active in all naval operations.'"[20]

Seamen risked the loss of their ships at least as much through their inability to ascertain the longitude as through enemy action. Defoe's novels indicate that on both counts those risks were considerable. But the event that caught public attention more than any other was Admiral Sir Cloudsley Shovel's loss of a squadron of ships and two thousand men off the Scilly Isles, after returning from the attempt on Toulon in 1707. Of the eight hundred men on his ship alone, no one was saved. By 1714 even the tardy British Admiralty had been sufficiently stirred to make their offer of a prize of twenty thousand pounds—equal to far more than a million dollars in today's money—for discovering a method to ascertain the longitude at sea.

THE NEW AWARENESS OF ACCURATE TIME MEASUREMENT

Though marine chronometers would not begin to affect navigation and map making until the latter part of the eighteenth century, the more accurate pendulum clocks and spring-balance watches were quickly affecting London life. What is of direct concern to this study is the way in which a sixty-fold improvement in the accuracy of clocks would affect time measurement itself. It had been discovered as early as the fifth century B.C., with the aid of a gnomon, that the length of the solar day, as measured by sundial time, was not uniform throughout the year.[21] However, for more than two thousand years thereafter, sundials and verge-and-foliot escapements were sufficiently inaccurate that the discrepancy remained of no particular consequence.

The discrepancy between the lengths of the solar day and the mean solar day is known as the "equation of time" and has two components. The first relates to the fact that the velocity of the earth's motion is not uniform, because the earth's orbit is an ellipse and not a circle. The velocity is greater when the earth is nearest the sun, in January, than when it is farthest away in July. This element in the equation of time adds nearly eight minutes to the day in April and subtracts approximately the same amount in October. The second component results from the fact that the sun's apparent motion is in a plane inclined to the equator. As a result, the component of the sun's velocity parallel to the equator is variable. The second component adds a maximum of ten minutes to the mean solar day in February and August, and reduces the day by ten minutes in May and November. Since the first and second components affect the total equation of time simultaneously, the net effect is for the solar day (apparent time) to be fifteen minutes longer than the mean solar day in February, and about seventeen minutes shorter than the mean solar day around November. This results in a maximum difference of about thirty-two minutes in the length of the day during the course of one calendar year.[22]

Derek Howse notes that, as a result of the increased accuracy of clocks, "communities began to keep mean time in preference to apparent time—Geneva from 1780, England from 1792, Berlin from 1810, Paris from 1816."[23] Individuals, however, had certainly been dealing with this adjustment for about a century before then. Most early longcase clocks were sold together with a sundial. This was because the sundial was still as essential for setting clocks as it had been in the earlier days of the verge-and-foliot escapement (figure 1). There were, however, several methods used to adjust the reading on the sundial in order to allow for the equation of time. The equation clock that Thomas Tompion, the father of English clockmaking, presented to the Pump Room, Bath, in 1709, was equipped with a rare equation dial. More commonly, equation-of-solar-time tables were printed on "watch papers" as a form of advertising. They were put in the back of the pair-case of any watch that had been sold or repaired. Such tables were also stuck on the inside of the door of a clock for ready reference. Joseph Williamson (d. 1725), a first-rate horologist and the maker and inventor of equation clocks, used the feature of a rise and fall of the acting length of the pendulum to adjust for the equation of time.[24]

A century earlier, John Donne, in his *Obsequies to the Lord Harrington*, suggests that Harrington has a soul which is so true and so closely regulated by God that it can both control the sun and act as a great "sun dyall." By this sundial all the other people and clocks might, in their turn, be regulated.[25] Ironically, the advent of pendulum clocks,

Figure 1. This copy of a mid-eighteenth-century painting by François Boucher, Paris, shows a young lady setting her watch at noon by a sundial. (Bayerisches Nationalmuseum, Munich)

after 1657, did not obviate the need to set one's clock by a sundial, but it did question solar time itself. Thus there came about "mean time" as yet another level of rationalization that followed upon the earlier introduction of equal hours. Until, in 1792, the state finally stepped in, each person tried to produce his own mean solar time by adjusting sundials to allow for the equation of time. This is not the only reason for Alexander Pope's famous couplet,

> 'Tis with our *Judgments* as our *Watches*, none
> Go just *alike*, yet each believes his own.

But the individual adjustments for the equation of time may be a contributing factor that does not seem to have been observed.[26]

Until 1657 most clocks—and this is a direct reflection of their inaccuracy—had only hour hands. But with the advent of the pendulum escapement it was not long before minutes and even seconds were indicated. Chaucer (d. 1400) nowhere mentions minutes or seconds in his canon, except as the sexagesimal divisions of degrees into minutes and seconds (second minutes), as used by astrologers and astronomers since Mesopotamian times. This more rationalized division for time would take over from Trevisa's irregular division of the hour, or the comparable table of Papias in Du Cange. As listed in the *Oxford English Dictionary* under *atom*, the hour in the table of Papias contained either 5 points, 10 minutes, 15 parts, 40 moments, 60 ostents, 480 ounces, or 22,560 atoms. It would be difficult to envisage such irregular divisions being indicated by a clock.

Like Chaucer, Shakespeare (1564–1616) had difficulty in denoting very small periods of time. Unlike Chaucer, however, Shakespeare already shares with his audience the concept of what a minute is like, and uses the term in our modern sense on more than sixty occasions. Yet I have nowhere found the term *second* in his canon used as a measurement for time. Nor, though hours are frequently mentioned, does Shakespeare any longer feel the need for specifying "equal" or "inequal" hours, as Chaucer does. Shakespeare's time is already on the way to being rationalized. By the end of the seventeenth century, individuals were taking account of the equation of time, and, by 1792, England instituted mean rather than apparent time as a rationalization of the solar day itself.

The next important rationalization of time grows directly out of the industrial revolution. Steam power contributed greatly to that revolution, and it did so in three stages: for pumping water out of mines (Savery, 1698; Newcomen, 1705; Watt, 1763 and 1769); for contributing to the mechanical power in factories during the industrial revolution (Watt, 1781, though water power was more important in the early stages); and for producing the revolution in transport during the first half of the nineteenth century. With the exception of our discussion of the use of chronometers at sea, the time with which we have thus far been concerned is local time. For example, when it is noon in Greenwich, it is 7:00 A.M. in New York. But even in Bristol, 112 miles west of Greenwich, local time is ten minutes behind Greenwich time. This had very little impact on human affairs when clocks were inaccurate and transport was slow. By the first half of the nineteenth century, however, the post office, the telegraph companies, and the railways were all feeling the urgent need for a further rationalization of time. The inevitable outcome would be the institution of a single time zone to cover the whole of Britain.

THE CONTINUING RATIONALIZATION OF TIME MEASUREMENT

Derek Howse lists some of the developments in communications between 1820 and 1850 that would affect both time and timekeeping: "the first public passenger train in 1825; the first Atlantic crossing under steam power in 1827; Wheatstone's electric telegraph of 1836; mail sent by rail from 1838; Bradshaw's railway timetables of 1839; Bain's electric clock of 1841; the first public telegraph of 1843 (running alongside the Great Western Railway line from Paddington to Slough)."[27] These developments were being pioneered in England. They resulted in immediate pressures for a further rationalization of time measurement. This was to be a uniform time system throughout England, rather than throughout the world. Clearly the post office, the telegraph companies, and the railways were operating with growing difficulty in a country increasingly frustrated by the lack of a single time zone.

In 1840 Captain Basil Hall, R.N., the explorer and one-time commissioner for longitude, wrote to Rowland Hill, who had introduced the penny post. Hall's proposal to Hill was that "only one expression of time would prevail over the country, and every clock and watch indicate by its hands the same hour and minute at the same moment of absolute time."[28] In November of that same year, the Great Western Railway ordered that London time should be kept at all its stations and in all its timetables. Other railways quickly followed suit, and Greenwich mean time soon became the *de facto* standard for the whole country. As is not unusual, the government proved particularly tardy in confirming the use of Greenwich mean time. The Statutes (Definition of Time) Bill did not receive royal assent until forty years later, on August 2, 1880. It stated, in part, that "whenever any expression of time occurs in any Acts of Parliament, deed, or other legal instrument, the time referred shall, unless it is otherwise specifically stated, be held in the case of Great Britain to be Greenwich mean time."[29]

By 1880, of course, other and more global changes in the rationalization of time were becoming increasingly necessary. These pressures would culminate in the International Meridian Conference of 1884, which recommended, among other things, that the universal time should be Greenwich time. The earlier problems of the British post office, telegraph companies, and railways were by then being enacted on a much larger scale. In 1830 seventy-three miles of railway track had been laid in the United States; by 1860 there were more than thirty thousand miles. In Canada, when British Columbia entered Confederation in 1871, the most important condition of union was the building of a railway to the east. By 1887 this had resulted in the Canadian Pacific, a genuinely transcontinental railway system extending from the Atlantic to the Pacific. Clearly, a global standard for time measurement had become essential. The principal proponents were Sir Sandford Fleming of Canada and Charles F. Dowd of the United States. As a result, standard time was instituted in 1884. This involved dividing the globe into twenty-four meridians that were fifteen degrees apart in longitude, starting from Greenwich. Each meridian was at the center of one of twenty-four time zones in which the time adopted would be uniform. This is by and large the system under which the world operates today.

Although the world may not then have realized it, the decision made in 1884 marks, perhaps as much as anything, the beginning of our global village. A corollary of the international agreements of 1884 was that the meridian passing through the Greenwich Observatory would be accepted as the standard for maps and shipping. The British had long used this as the zero or prime meridian. It coincides almost too neatly with our story of Britain's navy and empire that the Greenwich Observatory had been founded by Charles II—in 1675 and shortly after the Restoration—for the purpose of "perfecting navigation and astronomy." By 1950 the Greenwich meridian, which had earlier met with some resistance, was in almost universal use. Thus, in addition to standard time, there was a related and equally important rationalization, which provided universal map coordinates for the world as a whole. This rationalization—together with such forms of coordinating time as time balls, telegraphs, and radio—has increasingly rationalized both time distribution and marine and air navigation.

Nevertheless, the latter half of our century sees us pressing on apace with a further rationalization of time, which far exceeds the apparent needs or even understanding of modern men and women. The pressures now come from physicists and astronomers, on the one hand, and interplanetary navigators, on the other. In our Einsteinian century, we measure the size of the universe in billions of light years. Each light year represents the distance that light travels during a year at the rate of some $299{,}792.4 \pm .5$ kilometers per second. Though such distances are beyond normal comprehension, we are also concerned with measurements of time as minute as the chronon, which equals about one-billionth of a trillionth of a second. Since 1967 our second has been completely divorced from the macro-movements of the solar system. The atomic second is now defined as the duration of $9{,}192{,}631{,}770 \pm 20$ particular oscillations within a cesium 133 atom.

In addition, the present accuracy of atomic clocks and hydrogen masers has made us aware that the rotation of the earth is slowing down considerably. Some eighty-five million years ago, the solar year consisted of 370.3 days, while some six hundred million years ago, it equaled about 425 days. In order to correct our time system and coordinate the world's atomic clocks with the earth's rotation, leap seconds are now having to be added periodically to our clocks. Furthermore, there are a number of seasonal variations—probably related to wind patterns and tidal actions—that result in a maximum seasonal variation to the length of the day of about .5 milliseconds.[30]

Were it not that the millisecond is one-thousandth part of a second, one might be tempted to think of these variations as being comparable to the equation of time used for correcting sundials. It is now customary, however, to talk of clocks with an accuracy of one second in thirty thousand or even one hundred thousand years. At the Harvard-Smithsonian Center for Astrophysics there is a hydrogen maser, a microwave version of the light-generating laser, that is claimed to have an accuracy "equivalent to one second every fifty million years." With reference to the need for such clocks, and the almost incredible rationalization of time that they imply, it is argued that they "play a crucial role in measuring the barely perceptible but inexorable movements of the earth's crust, in navigating space probes through the swarm of moons circling Jupiter and Saturn, and in discerning the subtle structures of astronomical objects billions of light-years away."[31]

As we come toward the close of the twentieth century, it is difficult for us to comprehend a future in which time must seemingly be divided into unbelievably small parts. But it is equally difficult for us to understand how our forefathers apparently lived quite well with an hour that varied between forty and eighty minutes in length during the course of a year. The purpose of this study is to concentrate on the nature of the rationalization that has taken place over a very wide area of human endeavor during the past three hundred years. By studying past rationalizations, we may anticipate some of the future directions that rationalization will take. Such considerations are important for all of us.

Rationalization has made the Western world what it is today, a technocracy that is condemned and desired, envied and emulated. We shall find that, as we break each element of rationalization down into its component parts, there are strong reactions to change at almost every stage. Our main purpose will be to uncover as much of the rationalization as we can, but some space must be allocated to the resistance. If we ignore the challenge to change, we do so at our peril.

Since historical change builds on or evolves out of the past, the past itself provides one form of the challenge or reaction to change. Let us consider this phenomenon in terms of the rationalization of time. The division of the day has finally been established as twenty-four equal hours, probably because that evolves comfortably from an earlier day of two periods of twelve variable hours, or two Chinese and Japanese periods of six variable hours, or four tides, or six watches of four hours each, or eight watches of three hours each. Seemingly such pressures of history as the Mesopotamian sexagesimal number system have much to do with our acceptance of the day's being divided into twenty-four equal hours. The historical challenge to change has virtually eliminated the decimal system from diurnal time measurements, even though a day could very logically be divided into one hundred hours, each composed of one hundred minutes that contained one hundred seconds. And so, indeed, could the longitude be divided into a similar one hundred degrees, each composed of one hundred minutes that contained one hundred seconds.

At the 1884 international conference, both the time zones and the prime meridian were centered on Greenwich. At that time the seventh resolution was proposed by France, which has never willingly accepted the primacy of England. The resolution reads as follows: "That the Conference expresses the hope that the technical studies designed to regulate and extend the application of the decimal system to the division of angular space and of time shall be resumed, so as to permit the extension of this application to all cases in which it presents real advantages."[32] This resolution was put to the vote and was adopted unanimously. Nevertheless, the metric system has, until now, made only limited inroads into the measurement of time. International votes made mainly for the purposes of conciliation rarely do well against the forces of history.

Our present divisions of time clearly reflect the historical challenges to change. The day has been divided into twenty-four hours. But the hour has been divided by the sexagesimal system into minutes and seconds, which have enduring historical antecedents going back to the astrology and astronomy of Mesopotamia. Ironically, the new "minute" quantities into which we have quite recently divided the second have no historical antecedents. They, at least, have been accorded the level of rationalization that we associate with the decimal system. In descending order of time, these divisions of the second include milliseconds, microseconds, nanoseconds, picoseconds, and chronons, the last being given the name of time itself.

THE REACTION AGAINST CLOCKWORK VALUES

There are, of course, quite different ways in which man has reacted against the standardization of time and the related mechanization of his own nature. Dafydd ap Gwilym, the Welsh bard, is thought to have flourished between about 1340 and 1380, the very earliest time when mechanical clocks might have entered Britain. "A curse on its weights, a curse on its wheel," he says of the clock that has just woken him.[33] Though sundials sound no bell, they evidently could exert a comparable tyranny. Plautus (ca. 254–184 B.C.) demands that the gods confound the man

> Who in this place set up a sun-dial,
> To cut and hack my days so wretchedly
> Into small pieces!

Plautus continues,

> When I was a small boy,
> My belly was my sun-dial.[34]

Later, people complained in much the same way about clocks. In *English Proverbs*, for example, John Ray includes: "Your belly chimes, it's time to go to dinner." Eventually, poets reacted so strongly against clocks that they even looked back with nostalgia to the sandglass or sundial. William Blake, in *Jerusalem*, complains of the

> ... hour-glass contemnd because its simple workmanship
> Was like the workmanship of the plowman.[35]

In *On a Sundial*, William Hazlitt protests, "I never had a watch nor any other mode of keeping time in my possession"; and Charles Lamb, in *Old Benchers of the Inner Temple*, displays a similarly Romantic nostalgia for sundials, which involves the image that they have been man's friends all the way back to the time of the Garden of Eden.[36] Plautus would have turned in his grave!

But Romantic poets (and we are their heirs) did not just dislike the material values represented by an expensive watch. They sensed a more immediate threat from the order inherent in the clockwork urban society that surrounded them. During the British horological revolution, theologians and poets had agreed with philosophers that they lived in a clockwork Newtonian universe, whose God was an almighty watchmaker. Newton and Leibniz argued about details concerning the relationship between God and his clockwork; poets caviled that Descartes had not given his clockwork dog a soul that might protect it from vivisection, but by and large people celebrated the new horology. Samuel Richardson's Mr. B is "a regular piece of clock-work," and Grandison, his ideal man, is epitomized by the watch that he seems always to be carrying.[37] Benjamin Franklin reflects the values of his age when he coins the phrase "time is money."

After about 1760, however, attitudes toward clockwork values tend to change. Poets begin to see such values as being diabolical. Like the scientists in the century before Darwin, they upgrade the organic and downgrade the mechanical values. Only the bourgeoisie continued to esteem clockwork regularity; while theologians, similarly, retained the argument from design based on the clockwork universe. Since this is a subject on which I have written at length in *Clocks and the Cosmos: Time in Western Life and Thought*, I will not pursue it further here. Suffice it to say that the pejorative values that we now attach to acting like an automaton or being a mechanical man, rather than a "Man of Feeling," derive from our reaction to the rationalization of time. And that reaction grew directly out of the British horological revolution.

There is, of course, a considerable irony in our attitude toward clock values. We all desire the fruits of technology, yet we object loudly to the regimentation that technology requires. Fortunately, we have all witnessed a vociferous reaction to mechanical bourgeois values during the 1970s and the latter part of the 1960s. When this is considered carefully, the hippie movement bears a close resemblance to Romanticism on a large scale.[38] And, just as the few Romantics who lived to old age (Wordsworth, Hugo, Goethe) eventually accepted the comforts associated with bourgeois mechanical values, so the hippies of the seventies became the yuppies of the eighties. They may have a little more style than their bourgeois parents, but they have also become even more conventional, money-oriented, and self-centered than the people against whom they reacted.

Jonathan Swift—who, in his *Tale of a Tub*, described satire as "a sort of *Glass*, wherein Beholders do generally discover every body's Face but their Own"—must be laughing gleefully at the dilemma in which we now find ourselves. We worship abjectly the material advantages that the rationalization of time has brought to us. Yet we also complain loudly about the rationalization not only of production but also of ourselves, which is seemingly an unavoidable corollary. Were there ever such pointed horns to a dilemma? Swift's frenetic Gulliver, who represents the Moderns like ourselves, confides in us, "I had my self been a sort of projector [promoter and entrepreneur] in my younger days" (*Travels* 3.4). Swift leaves us in no doubt about the relationship between the bourgeois Gulliver and clockwork. The giant Brobdingnagians, who are expert in clockwork, think that Gulliver is himself "a piece of Clock-work," whereas Gulliver's watch is reported by a Lilliputian to be "the God that he worships."[39] Will we ever learn to see our own faces in Swift's glass of satire?

II Calendars
The Days of the Year

When we think of time, we normally think of the divisions of the day into hours, minutes, and seconds. These divisions, though they tend to be controlled by an evolution from historic precedents, are entirely arbitrary. When we think of the calendar, we think normally of the divisions of the year into seasons, months, weeks, and days. In this case, the divisions are by no means as entirely arbitrary as they may at first seem.

Though the day derives from the natural unit of light and darkness, primitive man tended to count his days as so many dawns, suns, nights, or sleeps. For example, the ancient Teutons used terms, derived from counting the nights, such as sennight (a week), and fortnight (two weeks). The week is an entirely artificial division of time, which seems frequently to have been the interval between market days. For example, among some West African tribes the interval was four days, in Central America it was five days, among ancient Assyrians it was six days, in ancient Rome it was eight days, and among the Incas and the ancient Egyptians it was a ten-day interval. Thanks to Christianity, the seven-day interval has prevailed. Perhaps not correctly, we think of it as deriving from the quarter of a lunar month. Certainly, the Israelites had a seven-day week, at least since the Babylonian exile of 586 B.C. In their case, however, the seventh day involved a taboo on trading.

In ancient times the Romans had the octonary week. This represented the eight days in which the farmers worked between one market day and the next, known as the *Nundinae* (from *novem dies* or "ninth day"). Later, as a result of both astrological and Judaic influences, the planetary or seven-day week took over. The seven planets—in the order Sun, Moon, Mars, Mercury, Jupiter, Venus, and Saturn—still provide the names for our days in either English or French. The seven-day week seems to have been taking over in Rome during the time of the Caesars. Its first public record there is found in a Sabine calendar of 19 B.C. to A.D. 14. By A.D. 200 the seven-day week was popular throughout the Mediterranean.

In view of Saturn's importance, however, Sunday did not become the first day of the week until the fourth century, when Mithraism replaced Saturn with the Sun.[1] In related developments, the pagan festival of the "Birthday of the Unconquered Sun" (*natalis solis invicti*) was inaugurated on astrological grounds by the Emperor Aurelian in A.D. 274. December 25 was assigned to the festival, because at that time, just after the solstice, the sun's light begins to increase. However, the Christians outmaneuvered the Mithraists. Christmas was celebrated in Rome from A.D. 336, on exactly the same date of December 25, which also became the first day of the church year.[2] Today, Christians still honor the Sun's day, but the Jews have remained faithful to Saturn's day.

Among primitive people the most important period, after the day, was the lunar month. This period is related to many natural phenomena, including tides and animal life, but it is not appropriate for calculating long portions of human existence. For shorter periods, however, it did well enough, and it became entrenched in the civilizations that have influenced the Western world. The Jewish calendar is still a lunar one, and the lunar calendar continues to be used by both Christian and Islamic theologians. It was also the system originally used in Rome.

Most particularly, of course, the lunar month is associated with female menstruation and its related taboos. Many examples might be cited, but one from Martin P. Nilsson will suffice: "The Samoan woman looks at the moon and expects the beginning of menstruation at a quite definite position of that planet, each woman naturally having a different position of the moon in view. If menstruation does not take place then, she perceives that she is pregnant, and expects her confinement after ten moon-months."[3] Months, moons, and menstruation are so closely related that we use the same root-word for all three of them.

THE EGYPTIAN USE OF THE SOLAR YEAR

While lunar months may be used for denoting relatively short periods, they are not by any means in cycle with the seasons. For agricultural people, the seasons were all-important. In tropical countries, time could be reckoned by the alternating rainy or dry periods, two of each in a year. But in more temperate regions the needs of sowing and reaping made increasingly important the development of a solar calendar to regulate the agricultural year. Perhaps for this reason—as well as because of special conditions related to the annual inundation of the Nile—the ancient Egyptians were early developers of the solar year. Like all other civilizations, however, they had to face the problem of the incompatibility of the lunar month and the solar year. We now know that the synodic month, which is the interval between two new moons, comprises on average 29 days, 12 hours, 44 minutes, 2.98 seconds. The solar year, during which the earth performs one revolution in its orbit around the sun, contains approximately 365 days, 5 hours, 48 minutes, 45 seconds of mean solar time. This so-called "tropical year" is 11 minutes and 15 seconds less than $365^{1/4}$ days.

At a very early date, the Egyptians noted that the heliacal rising of the star Sirius, or Sothis, corresponded very closely to the all-important rise of the Nile. They chose this for the first day of the year, and counted as one year the period between two such risings of the star. This was particularly convenient in that it was much longer than the month, and also included a complete cycle of seasons. In subdividing the new unit, the Egyptians typically superimposed it on the old division of the year into twelve months. Thus the year was divided into twelve nominal months of thirty days each, and the four-month periods contained respectively the three seasons of inundation, winter or sowing, and summer or harvest. Five further days were added on at the end of the year, making a total of 365 days. As we now know, however, the star year, which is virtually identical with the solar year, measures about $365^{1/4}$ days. As a result, Sothis rose approximately one day later every four years.

After 1,460 solar years (365 × 4), which has come to be known as a Sothic period, the Egyptian New Year's day was back in its right place again. The Latin writer Censorinus (fl. third century A.D.) notes, in his

De Dei natali, that the Egyptian New Year coincided with the rising of Sothis in A.D. 139. It has therefore been assumed that the original New Year must have been instituted in 1321 B.C., 2781 B.C., 4241 B.C., or 5701 B.C. Religious texts concerned with the pyramids of the fifth and sixth dynasties show that the calendar, with its five extra days, was then already in existence. As a result, Egyptologists choose either 4241 B.C. or 2781 B.C. for the introduction of the calendar. The date they choose reflects their views concerning the age of the pyramids. Though they rationalized their calendar so impressively and so much in advance of their contemporaries, the Egyptians have left us in doubt by an incredibly wide margin regarding the date when that calendar was introduced. But there is a very important reason why this was so. Because the Egyptians used no standard unit of time longer than a year, they could not date by eras or centuries in our modern sense. We shall look at this aspect of time later when we consider the rationalization that has taken place in chronology.

RATIONALIZATION THROUGH THE JULIAN AND GREGORIAN CALENDARS

Despite the embarrassing irregularity involved in the Sothic period, the rationalization of the Egyptian calendar was very impressive and in some respects has still not been surpassed. Julius Caesar, after his victory at Pharsalus in 48 B.C. had made him master of the known world, considered, while he was in Egypt, how to overcome the calendrical problems of his empire. The calendrical chaos in Rome was of no mean order. When the Julian calendar was instituted in 46 B.C., the year had to be stretched to 445 days. This was done in order to bring the existing lunar calendar into step with a solar cycle from which it had come to deviate by some three months.

It would be difficult to overemphasize the importance of the Julian calendar, not only for the Roman Empire but for Western civilization as a whole. The earliest Roman year—that of Romulus, the legendary founder of Rome—consisted of ten lunar months. These were thought of as a period equivalent to human gestation. The first two months of the year (which were subsequently January and February) were originally not included in the calendar because they were considered to be unproductive time. The original ten months were evidently designated by numbers, and six of them retained their places and numbers until the advent of the Julian calendar; these were Quintilis (later named for Julius), Sextilis (later named for Augustus), September, October, November, and December. The two months added to the earlier ten-month year were January (named for Janus, the two-headed god who looks both ways), and February (from *februare*, to purify).[4] At first, these were added as the last two months of the year, and only in 153 B.C. did January 1 become the first day of the civil year.

For his work on the calendar, Julius Caesar had the assistance of Sosigenes of Alexandria, the greatest Greek astronomer of that time. As elsewhere in antiquity, however, the Roman calendar was under the control of the priests, or pontiffs. Even a Caesar could not overthrow tradition completely. Julius had to operate within the framework of the existing calendar's unequal months and numerous feast days. As in so many other fields, even this rationalization had to pull its forelock to the powers of tradition. As a result, Julius took the old 355-day year and attached the remaining 10 days to the ends of appropriate months, arranging this so that the feast days were retained unchanged. He took good care, however, that the month which was named after himself, in 44 B.C., had 31 days. Furthermore, Julius failed to revise the unnecessarily complex numbering of the Roman calendar's days, which involved, among other things, counting back from the ides in the middle of the month.

Julius Caesar also created the first leap year by adding an extra day to the February of every fourth year. This simple intercalation meant that the Julian calendar would err by only one day in 128 years. But even such a seemingly effortless adjustment was too much for the college of pontiffs, who, quite incorrectly, added the extra day every three years. In order to rectify their error, Augustus, in 9 B.C., had to discontinue the intercalation for 16 years. He is reputed to have rewarded himself by appropriating the month of August. This involved taking a day from February, in order that his month might equal that of Julius.

Though the rationalization of the calendar by Julius Caesar involved a change of the greatest importance, with the passage of time it, too, needed to be rectified. In 1514 the pope's secretary asked Copernicus to look into the question of calendar reform. Copernicus felt it was essential that he first understand the relationship between the motions of the sun and the moon. On May 1, 1514, he circulated a manuscript called *The Little Commentary*. His fully developed heliocentric argument was made in *The Revolution of the Heavenly Spheres* (1543), from which we derive our modern meaning of the word *revolution*. But this work was, perhaps defensively, not published until the year of Copernicus's death. However, *The Little Commentary* had already suggested a sun-centered system with a moving earth. In it, therefore, Copernicus unavoidably questions the cosmologies of Aristotle (384–22 B.C.) and Ptolemy (second century A.D.), to which the church continued to subscribe.

Because they were seen as a convenient mathematical fiction, the Copernican views were not anathematized. In fact, the development of the mathematical figures for the new calendar involved both Protestants and Catholics. In 1551 Erasmus Rheingold, a Lutheran, drew up improved celestial tables based on the Copernican system; Pope Gregory XIII, who instituted the new calendar in 1582, was advised by the German Jesuit and mathematician Christopher Clavius (1537–1612). The essential change in the New Style or Gregorian calendar was that October 5, 1582, became October 15, 1582. In addition, no century year would thereafter be counted as a leap year until it was exactly divisible by 400.[5]

The Gregorian New Style calendar was not accepted by Great Britain for 170 years. When it was finally recognized, in 1752, it represented the last major rationalization of the Western calendar system. This is not to say, of course, that our present system is in any way ideal. The main difficulty is that, because of historical precedents, the months are of unequal length and are in no way related to the weeks. Numerous other systems have been proposed in attempts to rationalize the calendar. Many include blank days, as in the ancient Egyptian calendar. These, as occurs with the Aztecs, sometimes have sinister associations. The Baha'i faith, founded in the nineteenth century, uses a 365-day calendar composed of 19 weeks of 19 days each, plus 4 extra intercalary days. The Central American solar, 365-day calendar divides the year into 18 20-day months, plus 5 extra intercalary days. These are known among the Mayas as "days without names," and among the Aztecs as "hollow" or "superfluous" days.[6]

The Mayan calendar demonstrates a virtuosity in calendrical and chronological computation that is quite remarkable by any standards.

C. W. Ceram notes that the year according to the

Julian calendar is	365.250,000 days
Gregorian calendar is	365.242,500 days
Mayan calendar is	365.242,129 days
Siderial reckoning is	365.242,198 days

And yet the Mayan people, though able to make quite exact astronomical observations and handle a fairly complex mathematics, in other respects were in thrall to the worst form of mysticism. Having produced the world's best calendar, the otherwise rationalistic Mayas became its slaves.[7]

The sacred calendar of the Mayas was composed of 13 months of 20 days each, which represented a kind of symbolic alliance between human beings (associated with the number 20) and the 13 gods of the Upper World. As Georges Ifrah notes, however, "Since the Mayas used their 260-day sacred calendar concurrently with their 365-day civil calendar, the complete expression of a date required taking both calendars into account."[8] The civil calendar—which was of less importance to the Aztecs than to the Mayas—consisted of 18 months of 20 days each, plus the final 5 blank or unlucky days. The so-called Calendar Round—at which time a particular day of the sacred calendar coincided with that of the civil calendar—occurred every 18,980 days, which equaled 52 civil years or 73 sacred years.

As Ifrah notes, the Aztecs "believed that the end of each sacred cycle would be accompanied by all sorts of catastrophes, and as it approached they offered great numbers of human sacrifices in the hope of inducing their gods to let them live through another sacred cycle."[9] The Jews celebrated a comparable period, though it was not based on the same calendrical exactitude. Their Year of Jubilee—which was subsequently taken over by the Roman Catholic church as the Holy Year—followed Old Testament tradition (Leviticus 25:8–55). The fiftieth year—namely the year following seven sabbatical cycles of seven years each—was a year of rest. In this year, all hereditary properties were to revert to the original owners or their heirs, and all Israelite slaves, whose poverty had forced them into the employment of others, were to be emancipated (Exodus 21:2–11). The term has now been secularized as a celebration of a period of fifty or more years. An example was the Diamond Jubilee of Queen Victoria, in 1897.

There are many reasons why a more rationalized form of the calendar might have been instituted, if only to accommodate modern business and accounting practices. For example, the original week of various lengths seems to have been instituted to accord with the time between local market days. Today, however, our months are unequal, and those who are paid by the month receive the same remuneration for periods of unequal duration. Similarly, large institutions have to allocate larger budgets for leap years than for other years. Clearly a modified ancient Egyptian calendar would be much more convenient for aligning the days of the week with the months and the years. Such a calendar year might comprise twelve months, each containing three weeks of ten days, and a supplementary five or six days of Saturnalia. However, it is generally agreed that the religious associations of the seven-day week, ostensibly going all the way back to the creation in Genesis, have made it impossible to change that period, despite some very powerful advocates for reform.

The surprising tardiness of Great Britain in accepting the rationalization of the Gregorian calendar suggests, however, that other factors may also be involved. In 1582, the antagonism between Protestants and Catholics tended to ignore the fact that they were both sects of Christianity. Certainly the delays were all in the Protestant camp. Italy, Portugal, Spain, and France accepted the calendar in 1582, and the Catholic German states in 1583. It was 1700 before the Protestant German States followed suit. The British, however, did not conform until 1752, when they were obliged to omit eleven days following September 2. Like many other accounts, the excellent article in *Britannica* on the calendar reports on the misapprehension by the public. They apparently felt that they were beng cheated in some way, and there were said to have been related riots with the slogan, "Give us back our eleven days."

In a more recent article, Paul Alkon, who lists many of the reports, questions whether there actually were riots at the time and suggests that the riots might have occurred at the election held two years later.[10] Certainly, at the bottom right of Hogarth's print *An Election Entertainment* may be seen a flag on which is painted, "Give us Our Eleven Days" (figure 2). Even if there were no related riots in 1752, it is quite clear that there was a strong feeling against the change. Resistance

Figure 2. William Hogarth, *An Election Entertainment*, 1753–54. (Author's collection)

occurred despite the fact that England was then the world's most important trading nation. Those who are accustomed to dealing with British letters and documents of the period before 1752 are aware that the delay in accepting the Gregorian calendar frequently obliged writers to specify both the Old Style and the New Style dates when dealing with the Continent.

The date of the commencement of the New Year in England was also changed when the calendar was adjusted. Hitherto the year had begun on March 25, but this was now changed to January 1. The change had been enacted by the more rational Scots as early as 1600. Given their dependence on international trade, the delay of 170 years by the British is hard to explain. But perhaps great powers can become insensitive even to their own interests. How else are we to explain a comparable delay by the United States in instituting the metric system that had originally been supported by their own ministers? The subsequent changes in the Gregorian calendar have been infinitesimal. Since it was still in error by one day in 3,323 years, a further rule eliminates the intercalated day every four thousand years from the year A.D. 4000. As a result, the Gregorian calendar is now correct to within one day in 20,000 years.

Though it could no longer be regarded as a Catholic institution, Zerubavel argues that until the end of the nineteenth century the Gregorian calendar was observed exclusively in Christian cultures. By then "the push toward standardizing temporal reference at the global level was gradually gaining momentum. In 1873 and 1875 Japan and Egypt became the first non-Christian countries to adopt the Gregorian calendar." Both were also, at that time, beginning to modernize and Westernize. By "World War I, adopting the Gregorian calendar seemed to have become almost an obligatory ritual that *had* to accompany every revolution or proclamation of independence." This was true of Albania and China (1912), Estonia (1917), the Bolsheviks (1918), Yugoslavia (1919), and Turkey (1926).[11]

From the early part of our century, there has been an international movement toward the rationalization of the Gregorian calendar. This was probably because it had been accepted by Britain and the United States, and therefore epitomized modernization and Western technology. The powerful movement toward the Gregorian calendar, together with the religious authority of the seven-day week, may best explain why two influential calendars based on important revolutions (and offering superior rationalization) have failed dismally. This occurred despite the fact that the calendars of the Soviets and of the French Republicans successfully synchronized their weeks not only with the year but also with the months.

A REACTION AGAINST FURTHER RATIONALIZATION: THE SEVEN-DAY WEEK

The French Republican calendar of 1793 followed the ancient Egyptian 365-day calendar with its twelve months, each comprising three ten-day weeks. Obviously modeled after the five "epagomenal" or blank days of the Egyptian year were the five (or six in leap years) *sansculottides* of the French Republicans. Like the Central American, Baha'i, and Egyptian blank days, those of the French Republican calendar were all grouped together in a block at the end of the year. The French Republican calendar offered a continual synchrony between the week, the month, and the year. Thus Primidi (meaning "the first day") fell on the first, eleventh, and twenty-first days of every month; Duodi fell on the second, twelfth, and twenty-second days; and so on through Tridi, Quartidi, Quintidi, Sextidi, Septidi, Octidi, Nonidi, and Décadi. Logically enough, Décadi fell on the tenth, twentieth, and thirtieth of every month. The conversion of days and dates into one another was facilitated by the obvious affinity between the name of the day and the only three days in the month on which it could ever fall. For example, Quintidi (the fifth day) could only fall on the fifth, fifteenth, and twenty-fifth days of the month. Furthermore, the new names of the months suggested the seasons to which they belonged; and their suffixes, in groups of three, suggested to which of the four seasons each month appertained.

It would be hard to suggest a more complete form of rationalization than that achieved by the French Republican calendar. Given its obvious advantages in a world that was supposedly rational and moving fast in the direction of rationalization, how could it possibly have been rejected? For one thing Napoleon lost at Waterloo, and for another the French people themselves were not pleased with a system that involved working for nine rather than for six days before taking a day of rest. Also, the revolutionary calendar was too obviously republican and nationalistic. Calling the blank days *sansculottides*—in honor of the street revolutionaries—and the four-year cycle associated with leap years the *Françiade* would hardly endear the system to anyone outside France. Making the point of departure for the new Republican era September 22, 1792, was equally controversial.

The calendar reform of the French Republic was essentially an extension of their concurrent metric reform. Along with the new ten-day week, therefore, went a day now divided into ten hours. Each hour contained one hundred decimal minutes, and each minute one hundred decimal seconds. Since the "rational" and "scientific" decimal system was intended to de-Christianize France, it goes without saying that the saints' days were abolished. The most direct intention, of course, was the obliteration of the seven-day week, and the Christian Sabbath with which it had become so closely associated. As Zerubavel notes, "when the chief architect of the new calendar, Charles-Gilbert Romme, was asked what the main purpose of the new calendar was, he could reply unequivocally: 'To abolish Sunday.'"[12]

When dealing with other rationalized calendars, Zerubavel argues that "the failure of these calendars to gain official acceptance despite all this support can be explained only by a very deep societal resistance, which was explicitly articulated only by extreme Sabbatarians. . . . It was obviously only the prospect of the interruption of the continuous flow of the week by the introduction of 'blank' days that the Sabbatarians found objectionable about the World Calendar and the International Fixed Calendar." In this battle the Jews and the Christians had at last found common ground. Even the chief rabbi of the British Empire published a book-length report about the defeat of a related calendar proposal at the League of Nations. He hailed this as "a great victory in a fight for liberty second in importance to no other in many a century." In dealing with this subject, Zerubavel concludes perceptively: "While large parts of our sociocultural environment have been 'rationalized' during the past few centuries, the uninterrupted, continuous seven-day cycle remains to this day one of the most resilient 'irrational' cornerstones of modern civilization."[13] Although the French ultimately prevailed with their metric system, they failed absolutely insofar as temporal rationalization was concerned.

By closing his eyes to the Copernican heresy, for pragmatic reasons, Pope Gregory XIII had succeeded in putting the stamp of his church on an accurate though admittedly irrational calendar. It was rationalized only as far as it might be without eliminating the cycle of sabbaths. Also, the movable feasts, such as Easter, that were based on the old lunar calendar could still be fitted into the system. Perhaps just as important for the Western world was the less publicized genocide of saints' days, which was achieved by the Protestants. This change made a considerable contribution to the more regular calendrical rhythms conducive to bourgeois enterprise.

Even the followers of Karl Marx—who had branded religion as "the opium of the people" in his introduction to *Kritik der Hegelschen Rechtsphilosophie*—could not succeed in initiating a rationalized calendar. Lenin had been pragmatic enough to introduce the Gregorian calendar as early as 1918, less than three months after the Bolshevik Revolution, "for the purpose of being in harmony with all the civilized countries of the world."[14] In June 1929, after he had gained the interest of Joseph Stalin, Comrade Larin's proposal for an "uninterrupted production week" was examined by the "Rationalization Section" of the Supreme Economic Council. The new five-day week and thirty-day month, with five annual holidays interspersed throughout the year, was introduced on October 1, 1929.

The Russian system was naturally also anti-Sabbatarian, but it was far more radical than even the French Republican calendar. The intent—running parallel with the first "five-year plan," launched a year earlier—was to exploit industrial equipment continuously without allowing even nights or sabbaths to interfere. All the other systems permitted the possibility of rest days according to religious choice: Fridays for Muslims, Saturdays for Jews, or Sundays for Christians. Even the French Republican calendar had a rest day for the whole population on the tenth day. But the Russian system allocated a particular day of rest to a particular worker. Since on any given day eighty percent of the population would be working, this would seem to take rationalization of the calendar as far as it might go. Soon the five rest days came to be associated with particular colors: yellow for the first day, peach for the second, red for the third, purple for the fourth, and green for the fifth.

Locking people into particular color groups not only affected their societal relationships but even struck out at family life. It was not so much for this reason, however, as because of a consequent decline in productivity that the system was modified after two years. On June 23, 1931, Stalin singled out for his displeasure the lack of personal responsibility, which he claimed had resulted from the rotating workforce necessary for a continuous workweek. In December 1931 the system was modified to a week of six days in which the sixth day was regarded as a common day of rest. Since the calendar still used thirty-day months, the rest days were permanently set at the sixth, twelfth, eighteenth, twenty-fourth, and thirtieth days. Even so, many peasants took off both the official rest days and the traditional weekly days of worship. Zerubavel argues that: "As in France 140 years earlier, it was the essentially traditionalistic rural population who spear-headed the movement to preserve the seven-day week."[15] Eventually the authorities had to fix election days on official rest days that also coincided with Sunday in Christian areas, or Friday in Muslim republics. By doing so, in order to avoid even further work disruption on election days, they were implicitly acknowledging failure.

Nevertheless, the rationalized Soviet calendar lasted in one form or another for some eleven years. It did not finally come to an end until the Presidium of the Supreme Soviet restored the seven-day week on June 26, 1940. By that time, Russia had invaded both Poland and Finland. On June 22, 1941, it would itself be invaded by its "ally" Germany. Russia's new allies would be traditional Christians wedded to the Gregorian calendar. Since the state's best efforts had not enabled the Soviet Union to break the traditional seven-day religious rhythms of her own peasants, Stalin's return to the use of the Gregorian calendar was inevitable.

In the Russian and French revolutionary as in the Egyptian calendars, the five-, six-, or ten-day weeks fitted rationally into the thirty-day months. But in the Gregorian calendar there could be no rational relationship between the seven-day week and the months. The lengths of the months themselves were also irrational and gave rise to what Iona and Peter Opie describe as "the best-known mnemonic rhyme in the language—probably through its inclusion in the canons of the nursery":

> Thirty days hath September,
> April, June, and November;
> All the rest have thirty-one,
> Excepting February alone,
> And that has twenty-eight days clear
> And twenty-nine in each leap year.

The Opies give English versions of this mnemonic from as early as 1555 and 1577, as well as a thirteenth-century French version and a Latin version that may be even earlier.[16]

As we have seen with the resistance to dividing the day or its hours into decimal parts, tradition of any sort is hard to break. But it becomes harder when there is a direct relationship with religion. That is why even revolutionary leaders find it opportune to build on the religions of the past. Physical examples are the building of temples in Central America or churches in Europe (including the Holy Sepulchre) on sites that were sacred to previous religions. Frequently, too, both the dates and the customs of older festivals are adopted and adapted by the prevailing religion. For this, Christmas provides a particularly apposite example.[17] Christianity is not alone in being responsible for the persistence of the Sabbath and the festivals. The relationship between religion and the calendar goes back far beyond that.

If one ignores the rites related to hunting and gathering cultures and moves forward to the early agricultural settlements, one finds a close relationship between ruler, priests, and the setting of dates such as those for sowing and harvest. Because temples frequently faced the east and the rising sun, the terms *orientation* and *temple* are often almost synonymous. The figure of a god or goddess commonly faced the eastern entrance from within the darkness of the *cella*, or cellar. The temples (as in the Parthenon) were also oriented so that on a predetermined day the sun, as though by a miracle of the priests, would illuminate the cult figure. The same system appears in European burial mounds or barrows (ca. 2000 B.C. to A.D. 600). The "miracle" marks a precise time in the year from which other dates can be counted off. This ancient practice of counting off the days is reflected in the Jewish tradition of counting off the Omer of fifty days (seven weeks plus one day) from the second night of Passover until the night before Pentecost. Though there has never been a fully satisfactory explanation for Stonehenge, that it was used by priests for calendrical purposes is generally accepted.

Priests and festivals provide a form of order, which is based on a cycle of recurring events controlled by the calendar. The reason why we have not been able to rationalize the calendar fully—and in particular to eliminate the seven-day week—may have little to do with any inherent "magic" in the number seven. It is much more probably because Western technology has been developed by followers of the Judaic tradition. Jews, Christians, and Muslims now hold almost two-fifths of the world's population in the thrall of the seven-day week and its related Sabbaths. As we have seen, the structure of the week is not affected by the day on which the Sabbath falls. The Jews, who came first, chose the seventh day. This was the day on which God rested after creating the world. The Christians and Muslims, who came later, were therefore obliged to choose Sunday and Friday respectively. Jonathan Swift points up the nature of the choice with typical slyness when Gulliver complains, in *Travels* 3.2, that he is having to display himself to visiting giants "every day of the week (except Wednesday, which is their Sabbath)."

If the seven-day week of Jews, Christians, and Muslims had not prevented the Egyptian solar calendar from being fully transmitted to the heirs of modern technology, the tradition of twelve months of thirty days plus five or six blank days might well have come down to us. Within this framework there would have been the opportunity to rationalize the month further into weeks of ten, six, or five days each. But by the time that Julius Caesar attempted to take over the Egyptian calendar, the seven-day week was already too well established to incorporate the idea of twelve thirty-day months followed by five or six blank days. We shall find that virtually all forms of rationalization invite reaction of one sort or another. The reaction to the rationalization of the calendar has, however, been exceptionally strong. It would seem that the seven-day week, in particular, is destined to cast its "magic" spell over human beings for a long time to come.

III Chronology
The Years of the World

Chronology, the science of computing dates and arranging for their alignment with events, provides a method for measuring temporal distance from a fixed point or points in time. The chronologist's task is somewhat similar to that of the navigator at sea or in space who must use coordinates to fix his position. Both require that a fixed meridian or starting point be generally recognized—such as Greenwich, or the birth of Christ—and that the distance from that point be measured in agreed units. One discovers, further, that both the uses of the longitude at Greenwich and the use of chronological "slots" to organize historical events are surprisingly modern practices hinging on the development of more accurate clocks. We have already noted this with reference to navigation, but our present concern is with chronology.

Most of us are aware that the Greeks and Romans lacked our modern sense of chronology. But this was also true as recently as the Elizabethans, to which the clock that strikes three in Shakespeare's *Julius Caesar* (2.1.192–93), some fourteen hundred years before it was invented, bears witness. In fact, the term *century* did not have its chronological meaning until the first half of the seventeenth century. This is equally true of the term *progress*, in its modern sense of a progress through time rather than a royal progress through space. Similarly, the term *anachronism*, which presupposes an existing sense of chronology, was first used in 1646.[1] By 1650–54, however, Bishop James Ussher, in his *Annales Veteris et Novi Testamenti*, had even superimposed a modern form of chronology on the Bible. This use of chronology gained the support of both Sir Isaac Newton and Dr. Samuel Clark, among others. As a result, it was now possible to measure chronology from the precise moment at 9:00 A.M. on Sunday, October 23, 4004 B.C., when the creation occurred.[2]

The Bible had presumably always contained the necessary information from which its chronology was now derived. What had changed was that men were becoming sufficiently conscious of temporal distance to be dissatisfied with measuring it from genealogies, or dates related to the lives of their kings. They now wished to place both time and calendars within a precise all-encompassing framework of chronology. Eventually, of course, this would call into question the whole idea of a biblical creation. It would also bring an inevitable religious reaction to the subsequent scientific chronologies of the world and the cosmos. We must look first, however, at the extent to which Bishop Ussher's chronology represented a rationalization of what had come before.

CHRONOLOGY AND THE ESCHATOLOGIES

To put the matter in perspective, we must consider the question of eschatologies. The *Oxford English Dictionary* defines *eschatology* as "the department of theological science concerned with the last four things: death, judgement, heaven, and hell." Students of eschatology tend to divide the study of man's end into three main sections: naturalistic eschatologies, eternalistic eschatologies (both nonhistorical), and historical eschatologies. All the historical eschatologies derive ultimately from the Old Testament and involve mainly Judaism, Christianity, and Islam. Although under certain circumstances they might appeal to members of higher cultures, naturalistic eschatologies appear generally in primitive religions. They are limited, as to both time and space, in terms of the natural objects that their followers worship and the places that they know. Followers of such primitive religions seek to enjoy an organic unity with their immediate *Umwelt*.

When cultures involved in naturalistic eschatologies develop a sufficient level of spiritual sophistication, eternalistic eschatologies—such as those of Greece, the Orient, and China—are thought to grow out of them. Such eschatologies seem to be involved in an endless cycle of recurrence. All living things are caught up in the pain of this endless cycle. Their goal seems to be to escape from the incarnate, temporal world into the atemporal world of spiritual values. Greek myth, for example, involves annual cycles related to gods like Adonis and Persephone. But there are also the much longer periods—embodied in the succession of Uranus, Cronus, and Zeus—as brought together in the *Theogony* of Hesiod (fl. ca. 800 B.C.). Uranus, as sky, marries Gaia, as earth, and gives birth to Cronus, who later becomes the god of time. After Cronus castrates his father, with the aid of the sickle given to him by his mother, he sires Zeus, the god of day. And then, in his own turn, Zeus castrates the father and begins his own cycle. Hesiod describes the chronology, but he makes no attempt to provide precise dates. The same is true of the parallel Hurrian, Mesopotamian, and Anatolian theogonies. The great theorist of Taoism, Chuang-Tzu (d. ca. 300 B.C.), thought of the universe as a great current or flux in which states followed one another in a virtually uninterrupted, endless procession. Such eschatologies would not seem to have been conducive to the measurement of chronologies from a fixed point in time, like the creation of the earth or the incarnation of Christ.

The more highly developed eternalistic eschatologies did, however, require such complex mathematical tools as the Hindu-Arabic numerals, which have come down to us via the astrologers of Mesopotamia. The Hindu *kalpa*, or cosmic cycle in the "day of Brahma," involved, for example, a period of 4,320,000,000 years. Even this was no more than a small part of a series of unending cycles. Each "day of Brahma" contains one thousand smaller cycles or yugas, themselves divided into four aeons. The kalpas each involve successive world periods of four phases in which worlds emerge, flower, decline, and die. We are now, for example, in the fourth aeon of the Kali yuga, which began in 3102 B.C. and will continue for 400,000 years.[3]

Robert Silverberg quotes Maya periods as long as the *kinchiltun*, which had 23,040,000,000 days.[4] Such numbers can surely not be taken seriously in any modern historical sense. They are clearly intended to impress in something like the same way as the size of the Greek Titans, the 969 years of Methuselah, or the superhuman feats of Roland at Roncevaux.

The eternalistic eschatologies seem to be reaching out to simulate eternity through the use of large finite numbers. At a point of transition between the eternalistic and the historical eschatologies, the Zurvanites had twin gods: the bright god Ohrmazd was the god of infinite time, in contradistinction to the swarthy and evil Ahriman, who was the god of finite time or "Time of the long Dominion," a period of twelve thousand years. In a pact between Ohrmazd and Ahriman, their struggle was limited to four periods of three thousand years each. The later Zoroastrian cosmology also moved out of the eternalistic eschatology, and certainly influenced the Judaic cosmology after the Babylonian exile in 586 B.C. I have written about this elsewhere.[5]

Judaism, Christianity, and Islam are historical eschatologies. They differ from the nonhistorical eschatologies mainly by occupying a limited and linear historical time. Their story is concerned with a world of finite time, as set out in the Old Testament, New Testament, and Koran. The eternity promised with the houris of the Koran or in the heaven of Christianity (though not really in the Old Testament) can be expected in a very immediate future. Historical eschatologies deal with the relatively short linear time during which a single god works out the redemption of his people. The religions that derive from Judaism differ from it not only in their promise of heaven but also in their missionary orientation. That orientation—even more than the Holocaust, the pogroms, and the Inquisition—is why a Chosen People, which today has only some sixteen million followers, has engendered Christian and Muslim scions with a total population of between one and a half and two billion.

The historical eschatologies are involved with the story of people who were once alive. But there was no real attempt to place them in a historical context until the rise in time consciousness during the seventeenth century and thereafter. There were, of course, earlier endeavors to date events. Romans dated from the founding of their city in 753 B.C., Muslims from the Hegira of July 16, A.D. 622, and Jews from the creation of the world in 3761 B.C. But we can relate these dates chronologically only because we are prepared to accept the common denominator of the year in which Christ is presumed to have been born. Other reckonings are based on regnal years, dynastic eras, or counted from the date of a particular victory.

The disagreement between local calendars made it necessary to find a method of dating that could be widely understood. For this, the periodic Panhellenic festivals offered a common time standard. The numbering of Olympiads was introduced by either Timaeus (ca. 345–

ca. 250 B.C.), the historian, or Eratosthenes (ca. 275–ca. 195 B.C.), the head of the Alexandria library. The reckoning by Olympiads proved a valuable chronological tool for their successors, including Polybius. However, as E. J. Bickerman notes, "The trustworthiness of the earlier part of the list of Olympic victors, which begins in 776 BC, is doubtful."[6]

The current practice of measuring chronology from the incarnation of Christ seems to have come about almost by accident. In the year that we would now call A.D. 532, Dionysius Exiguus proposed a cycle of 532 years as an alternative to other such cycles of years for use in fixing Easter dates. Because he felt that one cycle of 532 years had just been completed since the birth of Christ, he used his system to work out some of the Easter dates for the second cycle. Others later extended the calculations. As W. M. O'Neil puts it: "In the West the system steadily won favour over others using other cycles."[7] In his *History of the Church in England*, the Venerable Bede (ca. 673–735 A.D.) employed the Dionysian years after the incarnation of Christ. In the second chapter he related an existing system of chronology with the incipient Christian system by stating that Julius Caesar's preparations for the invasion of England were made 693 years after the founding of Rome or 60 years before the incarnation of Christ. (Bede also wrote *De ratione temporum* in A.D. 725.) Later, Charlemagne made the practice of dating from the incarnation of Christ official, and it eventually spread to Eastern Christendom.

Despite the calculations of the Venerable Bede, it should be noted that, had the Romans registered the birth of Christ, they would not have used our system. Christ's date of birth would probably have been registered as the twenty-sixth year in the regnal period of Augustus, or the year in which Lucius Calpurnius Piso and Cossus Cornelius Lentulus were consuls—neither of which satisfies our modern sense of exact chronology. In fact, of course, our acute consciousness of chronology now makes us question Christ's birth date itself. This is because we know that the Herod who murdered the innocents died in 4 B.C.

However, as Zerubavel notes, the Christian—or Incarnation—Era spread slowly: to Catalonia in 1180, Aragon in 1350, Castile in 1383, and Portugal in 1422. From then on, the Christian Era began to proliferate. The main cause of this has been the extensive colonization by European Christians. Today, a date like "A.D. 1990" would mean the same in virtually every country of the world. This is a rationalization that has eliminated a whole host of local calculations of chronology. It has done so not because everybody believes that Christ was incarnated at a particular point in time, but because the convention is universally accepted for its secular convenience.[8]

LATITUDES OF CHRONOLOGICAL TIME

The acceptance of a single standard for chronology made people begin to think in terms of "latitudes of chronological time." Just as seamen knew that places might be found on a particular parallel or latitude of the earth, historians were beginning to recognize that different events had occurred during comparable parallels of time. Thomas Hearne (1678–1735) puts it this way in his *Ductor historicus*: "As Maps, by representing to our sight the Extent of Countries and the Distance and Situation of Towns, leave a clear and distinct Notion of them in the Imagination . . . just so do Chronological Tables figurate to us the Series and Concatenation of Times."[9] These parallels or latitudes of chronological time provided an exciting new perspective for the historian. Paul Alkon quotes from an essay on chronology in *The Preceptor*, which "observes that by using the Julian period to correlate events in Greece and Rome, it is possible to 'gather, that at the very Time *Alexander* was establishing the *Macedonian* Greatness in the *East*, an Empire was rising in the *West*, reserved by Providence to crush the Tyranny he was forcing upon Nations, at the Expence of so much Blood and Treasure.'"[10]

The similarity between parallels or latitudes in seamanship and in chronology was understood readily enough. In the period before the chronometer was available to determine the longitude, mariners used what was known as "parallel sailing" or "latitude sailing." In this method, the seaman first sailed to the latitude of his destination by using the relatively accurate procedure of determining his latitude at noon. He then sailed east or west along the parallel, until he arrived at the destination for which he was bound. I have shown elsewhere how, without feeling obliged to use the term, Defoe employs this method widely on both land and sea in *Robinson Crusoe II* (1720), *Captain Singleton* (1720), and *A New Voyage Round the World* (1725).[11]

Defoe also employs a corollary to parallel sailing and the related consciousness of latitudes in several of his projects for overseas settlement. These projects were based on the conviction that husbandry which is successful in a particular latitude will be equally successful if transferred to the same latitude in another part of the world. For example, Defoe projects cattle and crops similar to those found in England at the beginning and end of *A New Voyage Round the World*. He makes his recommendations for the east coast of South America between the latitudes of forty-two and forty-nine degrees south. These projects may also be compared with Defoe's extensive suggestions, in *A Plan of the English Commerce* (1728), for finding new areas in which to grow coffee, sugar cane, and tea. In each case, Defoe notes the latitudes in which these crops have been successfully grown in the past. He then suggests where future planting might be undertaken in similar latitudes elsewhere in the world.[12] Thomas Hearne, Defoe's contemporary, moves easily from the climate in parallel latitudes to the synchronism that may be derived from chronological tables: "As by a Map we may see the Whole Earth at once, and observe all the Countries that lie in this same Climate; so Chronological Tables give us a prospect of a general *Synchronism*, that is, the History of what has happen'd, and the eminent Men that have liv'd in the same Age, in all the several Nations of the World."[13]

Despite the increasing interest in chronology during the British horological revolution of 1660–1760, the many amateur chronologists of England continued to be frustrated by the retention of the Julian calendar. This remained part of an inadequate rationalization of chronologies. Defoe gives a somewhat facetious view of the current state of affairs in *The Political History of the Devil* (1726): "*Satan*, who, no doubt, would make a very good Chronologist, could settle every Epocha, correct every Kalendar, and bring all our Accounts of Time to a general Agreement; as well the *Grecian Olympiads*, the *Turkish Heghira*, the *Chinese* fictitious Account of the World's Duration, as our blind *Julian* and *Gregorian* Accounts, which have put the World, to this Day, into such Confusion, that we neither agree in our Holy-days or Working-days, Fasts or Feasts, nor keep the same Sabbaths in any Part of the same Globe."

Paul Alkon has shown how eighteenth-century English men and women were involved with calendars and with chronology to an extent that had not been possible before.[14] John Locke stresses the importance of teaching chronology in *Some Thoughts Concerning Education*

(1693). He argues there that it is "the fittest for a young Lad, who as soon as he is instructed in Chronology, and acquainted with the several *Epochs* in use in this part of the World, can reduce them to the *Julian Period*, should then have some *Latin History* put into his hand. . . . for wherever he begins, Chronology will keep it from Confusion."[15] This acute concern with chronology lies behind the eighteenth-century novels of Defoe, Richardson, Fielding and Austen, and makes them essentially a new and different genre. Although Defoe has been accused of anachronisms often enough, I have demonstrated elsewhere that they occurred not so much through slipshod work as because he was trying to satisfy the immediate interests of his readers while dealing with a protagonist's life of some seventy years. Furthermore, Defoe's anachronisms are of a different order from those of Shakespeare. In an article entitled "Clocks and Chronology in the Novels from Defoe to Austen," I have argued also that Richardson, Fielding, and Austen worked with an almanac in constructing their plots.[16]

Defoe's somewhat facetious suggestion that "*Satan* . . . would make a very good Chronologist" was to prove even more prescient than he can have imagined. The identification of Satan with finite time, as opposed to benevolent eternal time, is of long standing.[17] For Defoe to have identified him, therefore, with the new discipline of chronology would have seemed reasonable enough. The irony is that, during the seventeenth century, devout Christians had adopted chronology to give greater authenticity to the Bible. During the eighteenth century and beyond, however, this use of chronology would come back to haunt them. Chronology would then outdo Satan himself by calling into question the very essence of the biblical version of Creation. Using latitudes of chronological time to compare events in historical time was relatively safe. But now people began to apply their newly developed sense of time and progress to geological and biological developments. When that occurred, the rationalization of time measurement—to which members of the church had given their stamp of approval—would raise some alarming questions about the Bible that have still not been resolved.

In *The Political History of the Devil*, Defoe alludes to the fictitious ideas that the Chinese had about the duration of the world. Locke, writing three decades earlier, had referred to the same sensitive subject in his *Essay Concerning Human Understanding* (1689): "Some men imagine the duration of the world from its first existence to this present year 1689 to have been 5639 years . . . and others a great deal more; as . . . the *Chinese* now, who account the world 3,269,000 years old or more."[18] Defoe protects himself by pointing to the fictitious nature of the Chinese belief, and Locke tries to shock us by observing its preposterous enormousness. Yet the Chinese "fictitious Account of the World's Duration" is far less than one-tenth of one percent of our present estimate.

RATIONALIZING CHRONOLOGY AND THE REACTION OF THE CHURCH

Let us consider what the rationalization of the discipline of chronology has done for us during the past three hundred years. In the middle of the seventeenth century, James Ussher, archbishop of Armagh in Ireland, announced that God had created the earth in 4004 B.C. Dr. John Lightfoot, vice-chancellor of the University of Cambridge—who, eight years earlier, had published his own computations placing the date of creation at 3928 B.C.—applauded the accuracy of Ussher's work. Lightfoot then paid Ussher the compliment of refining the bishop's figures further. As a result, he was able to announce that man was created by the Trinity at 9:00 A.M. on October 23, 4004 B.C. So great was the faith in Ussher's theological learning, that in the next edition of the King James Bible his date of 4004 B.C. was actually inserted. It was entered in the margin of the section in which the creation is described. Other Ussher dates—such as 2394 B.C. for the flood, 1898 B.C. for the covenant of God with Abraham, and 1718 B.C. for the rise of Joseph to power in Egypt—were all entered in their appropriate locations. As Robert Silverberg puts it, "Soon Ussher's dates were looked upon by Christians as being as sacred as the Biblical text itself. To question them meant questioning the whole fabric of religion. It was a subversive act, pure heresy, to insist that the world was any older than Ussher said it was."[19]

Bishop Ussher, however, had engaged Christianity to chronology at the very moment when science and technology, spearheaded by horology, were beginning to progress at an exponential rate. Time would, thereafter, be measured in dimensions that were increasingly both smaller and larger than had ever been scientifically quantified before.

Thus the eighteenth century witnessed, on the one hand, both chronometers and stopwatches, but, on the other hand, the beginnings of geological time. In the nineteenth century came systems for virtually simultaneous time through telegraphs, electric clocks, and time zones, but also the progress through time involved in the evolution of all living things. The twentieth century has seen atomic clocks and hydrogen masers, which measure time to an accuracy of far less than a second in one hundred thousand years. But it has also introduced us to the heretofore unimaginable progress of a universe that has been expanding outward for billions of years.

The six thousand years or less since the creation, as envisioned by Bishop Ussher and his colleagues, has long seemed incompatible with the expanding chronologies of more recent disciplines. And the number and nature of these disciplines have also been expanding. Inherent in many of our modern disciplines is the model of progress based on a heightened awareness of time and chronology.

The new approach related to a progress through time permeates the philosophies and disciplines of the modern world. In the eighteenth century, Adam Smith's argument in *Wealth of Nations* (1776) that wealth will increase without government interference; Buffon's geological stages in *Epoques de la nature* (1779); and Condorcet's moral and social progress in ten epochs in *Tableau historique des progrès de l'esprit humain* (1795) are all equally dependent on the idea of progress and growth through time. So, in the nineteenth century, are the Hegelian dialectic; the Marxian dialectic; the Darwinian survival of the fittest; Thomsen's *Guide to Scandinavian Antiquities* (1836), which introduced us to the progressive ages of stone, copper, and iron; and Morgan's *Human Progress from Savagery Through Barbarism into Civilization* (1877), which would inspire the new discipline of anthropology. When Engels discusses the relationship between the dialectics of Hegel and Marx in *Ludwig Feuerbach*, he relates this directly to the idea of progress, which by his time had "thoroughly permeated ordinary consciousness."[20] The complementary idea of progressive strata runs equally through William Smith's *Strata Identified by Organized Fossils* (1814), which led to our system of geological ages based on strata; Lyell's *Principles of Geology* (1830), which described inorganic physical processes and influenced Darwin; Schliemann's digging in

Troy, which gave us archaeology; and Freud's digging in the mind, which gave us psychoanalysis.

In our own century, the Einsteinian expanding universe flows directly from the idea of progress through time, which is the basis of so many of our modern disciplines. But long before the age of the present model of the universe had been expanded to measure between ten and twenty billion years, the Christians had been reacting strongly to such heretical chronologies. These chronologies could not avoid questioning the authenticity of the biblical account of the creation.

The concept of a carefully documented progress through time had fitted less well with the civilizations before our modern Western age. The Chinese, for example, had a long tradition of histories written by bureaucrats for bureaucrats. These were composed for the purpose of extolling one's ancestors. Among Muslims a sense of progress could hardly be envisaged. Mohammed had instructed his flock through a Koran that was to be neither questioned nor changed in the light of any future developments. Much the same was true of Christianity throughout the Middle Ages. Such attitudes accorded well with a feudal society, and a period when chronology had not yet been extensively rationalized.

However, the wide acceptance of a single chronology based on the supposed date of the incarnation of Christ was bound to produce a Bishop Ussher; the increasing concern with finite time made it inevitable that one would want to know the precise moment of creation. Nevertheless, the experience of the last three hundred years has now made Bible societies a little more reticent about including that first date. The attraction of inserting some dates can, however, be hard to resist. In an age fascinated by time and chronology, the use of dates can give a powerful authenticity to the text. The Authorized Version of the Bible—published by the British and Foreign Bible Society in 1954 to celebrate their third jubilee—ventures no date before Exodus. The second book of the Pentateuch is modestly headed "B.C. ?1300–1200?" Thereafter, there are such dates as "B.C. 1000–900" for the City of David in 1 Chronicles, and "B.C. 600–500" for the Babylonian captivity.

The scientific investigations that led to the inevitable questioning of the biblical account of the creation began almost concurrently with

Bishop Ussher's formulation of its chronology. Nicholas Steno (1638–86), who took one of the first steps that led to a questioning of Bishop Ussher's chronology, was the son of a Protestant goldsmith in Copenhagen. By the age of thirty he had become a Catholic and had published a pioneering book on geology. This was generally known as *The Prodromus*, from its Latin title. Steno—who had studied medicine and anatomy, and based his work on a study of geology in Tuscany—recognized that fossils were the remains of organisms, and that strata resulted from sediment laid down at the bottom of seas or rivers. He understood the nature of volcanoes and the fact that mountains were shaped by running water. He also rejected the widely held idea that mountains grew. Instead, Steno attributed some mountains to "exhalations," and he maintained that strata had been tilted as a result of the collapse of great caverns. Most important of all, he provided elementary definitions of morpheous, igneous, and sedimentary rock; and he recognized that the older strata would normally be below those which had been subsequently laid down. Clearly, this short but revolutionary work had broken out from the static world-view based on a creation without a subsequent evolution through time. It would be hard, indeed, to produce the chronology for the world of Steno's *Prodromus* and keep it within the limits of Ussher's six thousand years.

The seminal nature of Steno's *Prodromus* was recognized by Henry Oldenburg, a great facilitator for science before and during the early days of the Royal Society. Oldenburg became one of the two first secretaries of the Royal Society in 1663, and he corresponded extensively with Spinoza. He had translated Steno's *Prodromus* into English by 1671. But Steno's own story is indicative of what could happen to anyone whose studies might imply a questioning of the official Christian chronology of the earth. In 1665 he had gone to Florence and been appointed physician to the Grand Duke Ferdinand II, and by 1667 he had become a Catholic. The publication of *Prodromus* (1669) had brought sufficient fame for him to be recalled home as physician to the king of Denmark and professor of anatomy in Copenhagen. However, Steno's need for eternal salvation made him turn his back soon enough on the study of science and its concern with finite time. He returned to Florence, where he was consecrated as a priest in 1675. Within a year Pope Innocent XI had made this valuable convert a bishop. He was appointed apostolic vicar of northern Germany and Scandinavia, with particular responsibility for organizing Catholic propaganda throughout northern Europe. During his later stormy career in Germany, Steno even engaged in an abortive attempt to convert Spinoza.

Steno went further than most in his frenetic pursuit of eternal salvation. However, the need for early scientists to recant when confronted by the church's apparent ability to damn them for all eternity is understandable enough. Galileo's recantation, as a Catholic, of his belief that the earth moved around the sun may well have stimulated the efflorescence of science in Protestant Europe. One can understand the recantation of the medieval poet Chaucer, before his death in 1400, but Descartes did much the same almost exactly a quarter of a millennium later. The advances in science made by Englishmen should not blind us to their need for great circumspection regarding religion. As R. F. Jones puts it in *Ancients and Moderns*, Francis Bacon's "emphatic separation of science and religion certainly made scientific progress easier and enabled the Puritans to embrace his philosophy."[21] At the end of the seventeenth century, John Locke shows a comparable obeisance toward religion in the "Epistle to the Reader," at the beginning of his *Essay Concerning Human Understanding* (1689). There, five or six friends have decided that, before examining some perplexing doubts, "it was necessary to examine our own abilities and see what objects our understandings were, or were not, fitted to deal with."[22] A century later, Immanuel Kant, in an even more restrictive society, put aside any question related to "God, freedom, and immortality," before engaging in his *Critique of Pure Reason* (1781).

Even Georges Louis Leclerc, comte de Buffon (1707–88)—who made the first modern attempt to embrace all scientific knowledge in his comprehensive work on natural history—could not ignore the power of the church. In his famous *Epoques de la nature* (1779), one of the forty-four volumes of the *Histoire naturelle* (1749–1804), Buffon outlined the successive geological stages. He based this aspect of his study on Newton's speculation that if our earth had been broken off from the sun, a globe of this size would have taken 50,000 years to cool. As a result, Buffon calculated to his own satisfaction that the age of the earth must now be 74,832 years. He conveniently divided the development of the earth into seven epochs, during only the last of which men had appeared. (This accorded suitably with Condorcet's later *Tableau historique des progrès de l'esprit humain* [1795], which deals in ten epochs with the evolutionary nature of man's social and moral progress.) However—despite Buffon's status as the keeper of the Jardin du Roi from the age of thirty-five, and his valuable knowledge of timber for the shipbuilding projects of the French government—he was by no means exempted from interrogation by church-related authorities.

As Daniel J. Boorstin reports, Buffon's first volume of the *Histoire naturelle* was investigated by a committee of the theology faculty of the University of Paris in 1749. Buffon wrote to the committee, "I abandon whatever in my book concerns the formation of the earth, and in general all that might be contrary to the narration of Moses, having presented my hypothesis on the formation of the planets only as a pure philosophical speculation." As a result of "kissing the rod," Buffon was delighted to discover that the committee voted by 115 to 5 not to censure the work. As Boorstin notes further, whether from piety or prudence, Buffon refused to be embroiled in theological controversy and sought the last rites of the church at his death. In 1773 he explained: "I do not understand theology and I have always abstained from discussing it."[23] As occurred with the earlier Robert Boyle, the father of chemistry, and Isaac Newton, the father of the mechanical universe, there would continue to be many who, like Buffon, apparently succeeded in balancing the claims of science and religion. After Buffon, however, it would become increasingly difficult to limit the years since the creation of the world to a mere six thousand.

SCIENCE AND THE EXPANSION OF CHRONOLOGY

In 1785 the Scottish geologist James Hutton (1726–97) dealt a further blow to the ideas of Bishop Ussher, when he published his *Theory of the Earth; or, An Investigation of the Laws Observable in the Composition, Dissolution and Restoration of Land upon the Globe*. Hutton held that the key to geology was slow and gradual change. Although he could not prove it, his "Law of Uniformitarianism" assumed that the geological processes, such as erosion and sedimentation, had been proceeding at a comparable rate since the earth's formation. Thus one could examine a layer of sedimentary rock and decide how long it had taken to form that layer. One achieved this by knowing the annual rate at which similar sediments were currently being deposited.

Not long after Hutton's *Theory of the Earth*, William Smith, an English canal surveyor, became involved with strata. His interest developed when he carried out an underground survey for coal in Somerset, and he produced early geological maps related to his canal surveys. Smith drew up a table of thirty-two different strata and listed the characteristic fossils for each. He noted that one could determine the relative age of a stratum by its respective fossils. In 1814 Smith published *Strata Identified by Organized Fossils*, and his study made it quite clear that many fossil strata contained fossil animals that were no longer in existence. By the mid-1830s Smith and his successors were producing a calendar of the past. They showed that the chronology of the world was divided into four great eras, based on differences in strata, and that human beings have lived only in the last or Quaternary era. Each era was further subdivided, and the periods named for where the key rock formations had been studied, such as the Cambrian in Wales. By way of exception, the Cretaceous period was named after the Latin word for the chalk in which the key fossils were found. This stratigraphic discovery, that similar fossils were generally being found in corresponding strata, was also developed at the same time by the French zoologists Georges Cuvier and Alexandre Brongniart.

Quite clearly, the studies in paleontology would influence Darwin. But Darwin's chief praise of a geologist was reserved for Sir Charles Lyell (1797–1875) and his *Principles of Geology* (3 volumes, 1830). On the plain of Catania, Lyell had found limestone strata containing marine fossils. These not only were similar to present-day organisms but passed right under the main volcano of Mount Etna. He was convinced that the mountain—which was now ten thousand feet high and ninety miles wide at the base—must have been formed over millions of years. It followed that the earth itself, as well as the many strata of sedimentary rock containing very different fossils, must be of an age beyond the possibility of measurement.

Darwin read the first volume of Lyell's *Principles* before leaving on the five-year voyage of the Beagle, on December 27, 1831. In the *Origin of the Species* (1859) he wrote: "He who can read Sir Charles Lyell's grand work on the Principles of Geology [will recognize] how incomprehensively vast have been the past periods of time."[24] In his earlier *Journal of Researches* (1839), which reports on the voyage of the Beagle, Darwin includes many references to Lyell, and all are in a tone of reverential praise. Geology in general, and paleontology in particular, had provided the vastly extended chronology that furnished an essential framework for his *Origin of the Species*.

The earlier biologists could operate without calling into question the temporal limitations that had been set up by Bishop Ussher. John Ray (1606–1705) was the first botanist who attempted to define what constitutes species. His *Historia plantarum* (1686–1704) described and classified all known plants. Ray's zoological works were characterized by Georges Cuvier (1769–1832), the French zoologist turned paleontologist, as "the basis of all modern zoology." But it was Carolus Linnaeus (1707–78), the Swedish botanist, who was surely the most prolific of all classifiers. In all, he produced more than 180 works. These were not limited to his main interest in plants and animals, but classified things as different as minerals and diseases. The work of classifiers like Ray and Linnaeus provided an essential underpinning for the increasing interest in biology. In a parallel development, which may well reflect the interest in biology, scientists and philosophers tended to use organic rather than mechanical models from about 1760. The change of interest from mechanical to organic models is also reflected in the observation by William Powell Jones, in the *Rhetoric of Science*, that "after 1760 the most popular subject for scientific study was natural history, especially botany."[25]

The imagery in Edward Young's *Conjectures on Original Composition* (1759) reflects the increasing appreciation of the organic rather than the mechanical model. When he defines what he means by "an *Original*," Young clearly recognizes the mechanical and botanical analogies for literature. Writing at the close of the horological revolution, his own preferences are clear: "An *Original* may be said to be of a *vegetable* nature; it rises spontaneously from the vital roots of genius; it *grows*, it is not *made*: Imitations are often a sort of manufacture wrought up by those *mechanics*, *art*, and *labour*, out of pre-existent materials not their own."[26] This foreshadows Coleridge's organic theory of imagination, as contrasted with the mechanical processes of the memory and fancy, which he downgrades. The essential elements for Charles Darwin's theory of evolution grow directly out of developments related to the increasing predilection for the organic model, after the latter part of the eighteenth century.

By 1779 Buffon's *Epoques de la nature* had begun to expand the temporal chronology of the earth. In the same year Hume's *Dialogues Con-*

cerning Natural Religion provided a powerful endorsement for the exclusion of the mechanical metaphor (of which Newton's clockwork universe was only one example) in favor of the organic one. In 1794–96 Erasmus Darwin, the grandfather of Charles, published *Zoonomia*, which contained a chapter on evolution. But Erasmus lacked a theory for what might have caused evolution to occur. Unwittingly, this was provided by Thomas Robert Malthus. Two years after the elder Darwin's *Zoonomia*, Malthus published his *Essay on the Principle of Population* (1798). The thesis of Malthus is stated most succinctly in the first sentence of his second chapter: "I said that population, when unchecked, increased in a geometrical ratio, and subsistence for man in an arithmetical ratio."[27]

Charles Darwin expanded the pessimistic Malthusian thesis concerning man to address all forms of life. His theory of evolution followed the Malthusian concept that only the fittest can be expected to survive. Alfred Russel Wallace (1823–1913), who produced his theory of evolution independently of Charles Darwin, also used the Malthusian thesis. It is ironic that in developing a theory that would eventually relate man to nature—rather than to God—both Darwin and Wallace borrowed from a thesis that was concerned solely with the human population.

CHRONOLOGY AND THE NEW DISCIPLINES BASED ON PROGRESS THROUGH TIME

The Malthusian thesis is grounded in demography. This is yet another of the new disciplines that began in the Restoration and eighteenth century, and are based on the concept of a development through time. Demography, or the scientific study of developments related to population, began in the seventeenth century with John Graunt, a London haberdasher. As a result of related experiments and observations, he and his friends received a charter as the Royal Philosophical Society, the first scientific academy in the modern world. Using the periodic records of christenings and burials issued by parish clerks, together with reports on the apparent causes of death, Graunt discovered many significant regularities. He published his findings in a pamphlet entitled *Natural and Political Observations . . . Made upon the Bills of Mortality of London* (1662). Graunt particularly attracted the attention of scientists by a hypothetical life table he included in the pamphlet. This showed the proportion of people born alive who might expect to live to successive ages. He also indicated the average expected years of future life for a person at each age.

In 1693—after Leibniz had forwarded to the Royal Philosophical Society of London information from the registers of Breslau, Germany, for the years 1687–91—Edmund Halley, the astronomer and friend of Newton, used these figures to construct the first empirical life table. Premium rates for life insurance could now be instituted; they varied with age and were at first based on the Breslau figures. Later a mortality table was constructed, which was based on the London bills of mortality for 1728–37. Daniel Defoe—a London tradesman like Graunt, but one who developed no mean knack for writing—used the London mortality bills for 1665 to produce the first novel based on statistics, *A Journal of the Plague Year* (1721). As one of the first successful journalists, Defoe knew all too well how the mortality tables from the period of the Great Plague might capture the popular imagination. He produced his novel at the precise time when another outbreak was expected to arrive from the Continent. Throughout the eighteenth century, the use of vital statistics continued. Out of this developed the census, which began in France and England in 1801. From the census of 1801, Malthus felt that he found support for his theory, as a result of the unexpectedly large increase in the population. When we consider the rationalization of human beings themselves during the nineteenth and twentieth centuries, we shall look again at some of the surprising uses that were developed for statistics and demography.

The theory of the survival of the fittest relied heavily, as we have seen, on the modern concept of progress through time. And the two main disciplines to which Darwin looked for support—Lyell's paleontology and Malthus's demography—also belonged to the new disciplines based on the concept of development through time. But the effect of *Origin of the Species* was more immediate. The public could not help but see the relationship between themselves and the flora and fauna. What was being questioned was the biblical chronology and the possibility that God could have directly created man. The reaction demanded that the Bible must be believed in its entirety or not at all. By 1864 eleven thousand Anglican clergy had signed the Oxford Declaration, which supported this black-or-white view. The battle lines were soon drawn. Though scientists, like Lyell and Huxley, supported Darwin, Bishop Wilberforce attacked him unmitigatingly.

Despite the attacks from some quarters, Darwin's theory had a compulsion about it that made others employ the "survival of the fittest" to support their own views, however bizarre. Naturally, Darwin's theory accorded most readily with other disciplines, such as the dialectics of Marx and Hegel, based on survival through time. Thus Marx wrote to Engels, "*Origin* is the natural history foundation for our views."[28] In Germany, Ernst Haeckel used the survival of the fittest to supplement Hegel's dialectic that celebrated the modern German nation-state. His argument for the development of a German superrace would lead directly to Nazism.

James Burke argues that our present world demonstrates the opposition between two forms of the Darwinian cult of progress: the Russian view of progress through planning and the American view of progress based on Herbert Spencer's social Darwinism of free enterprise.[29] Spencer's *Social Statics* (1851) argues the case for laissez-faire. He takes a Malthusian view about the pressures of population but gives this an optimistic slant, which naturally endeared him to industrialists. Carnegie and Rockefeller were admirers, and, in 1882, Spencer made a triumphant visit to America. As he had argued thirty years earlier in *Social Statics*, "progress is not an accident but a necessity."

In England first the Anglicans and later the Nonconformists have come to accept that the biblical chronology for the creation must be considered allegorical rather than absolute. Some fundamentalist sects still attempt to insist that the creation took place six thousand years ago, and in six days. Occasionally they modify the chronology slightly by claiming for God the qualities inherent in the words of the Ninetieth Psalm: "For a thousand years in thy sight are but as yesterday when it is past, and as a watch in the night." After Pius XII published *Humani generis* in 1951, however, even the Catholic church has permitted the discussion of evolution.

In July 1925, in Dayton, Tennessee, the teacher John T. Scopes was prosecuted for teaching the Darwinian theory contrary to the state statute. The relevant section of the statute stated that it was "unlawful for any teacher in any of the universities, normals and other public schools of the state, to teach any theory that denies the story of the divine creation of man as taught in the Bible, and teach instead that man has descended from a lower order of animals." The state of Tennessee had been the first to pass an antievolution law, on March 13, 1925. In spite of the fact that Scopes was cleared on a technicality two years later, the statute remained on the books until 1967.

Though Ussher's chronology of the Bible seems to have become less stressed, even during this author's lifetime, there are still attempts by Fundamentalist Christians to ensure that evolution should be presented as no more than a hypothesis. The possibility that we might have evolved from monkeys does, however, continue to remain a particularly sore point with some people. As a result of more modern methods for measuring the age of rocks, geology has once again been able to come to the aid of Darwinian evolution. In the 1930s, even before the confirmation of radioactive dating, geologists and paleontologists were generally agreed on the dates of past eras. They placed the beginnings of the Cenozoic era of mammals at some eighty million years ago, the Mesozoic era of dinosaurs some one hundred million years earlier, and the Paleozoic era at about four hundred million years earlier still. As Robert Silverberg puts it, "When radioactive dating began to supply a flood of figures based on the unvarying, unarguable ticking of the atomic clock, they proved to correspond quite well with the estimates that had been reached in a more intuitive way."[30]

There is nothing static about the standards by which we measure time. All that can be said at this stage is that the reaction of the church in relation to the calendar has succeeded in forcing the retention of the seven-day week. In that respect they have been able to hold back rationalization. In the matter of chronology, however, they seem to have lost the battle. The church's understandable interest forced them to support the incarnation of Christ as the international benchmark for rationalizing chronology. This led ineluctably to Bishop Ussher's chronology, which measured time from the biblical creation of the world. But the same interest in time that led to Ussher's chronology would also lead to a host of new disciplines based on a sense of development through time. These disciplines—and in particular geology and biology—would explode any possibility that the world was created six thousand years ago in six days.

In our own century there are new models of an Einsteinian universe expanding outward through time, which have greatly increased any possible age for the universe as a whole. We cannot, however, consider that expanding universe until we first deal with the rationalization of numbers, and particularly with that aspect of mathematics which is concerned with movement through time.

II Rationalizing Measures and Communication

IV The Ascendancy of the Metric System

The rationalization of length, area, weight, and volume has been perhaps even more extensive than the rationalization of time, calendar, and chronology. The *Encyclopaedia Britannica*'s article on "Weights and Measures" lists more than three hundred terms for weights and measures that are used in modern times. These are as varied as the *barile* of Rome (58.34 litres), the *bat* of Thailand (15 grams), or the *batman*. The *batman* for Iran is given as "6½ lb. av.; varies locally," and for Turkestan as "125 kg. (variable)." Although an attempt is made to indicate some ancient weights and measures, this is clearly a subject of even greater diversity.

Despite the very real endeavors that authorities have made to control them, weights and measures have for most of their existence involved a very inexact science. Injunctions against the use of false weights and measures ring down to us through the ages. As early as Deuteronomy 25:13–16, the Children of Israel were enjoined, "Thou shalt not have in thy bag divers weights, a great and a small. Thou shalt not have in thine house divers measures, a great and a small. But thou shalt have a perfect and just weight, a perfect and just measure shalt thou have: that thy days may be lengthened in the land which the LORD thy God giveth thee." It should not go unnoticed that the reward promised for following this seemingly bourgeois injunction is virtually identical with that promised for honoring one's father and mother in Exodus 20:12 and Deuteronomy 6:16. But the punishment promised for dealing with false weights is devastating: "For all that do such things, and all that do unrighteously, are an abomination unto the LORD thy God."

Although we know well enough that the measures can hardly have been accurate by modern standards, the Bible has many references to weights and measures, and is frequently very specific. In Genesis 7:15–16 the size of the ark is specified as three hundred cubits in length, fifty cubits in breadth, and thirty cubits in height. Since Noah was instructed to carry in the ark "of every living thing of all flesh, two of every sort," we may assume that they were even more tightly packed than the black slaves on the middle passage. In 1 Samuel 17:4–7, the size of Goliath and the weight of his armor are carefully specified for the purpose of impressing us. And in 1 Kings 6 and 7, the very careful setting down of the measurements and related specifications stresses the importance placed on "the house which king Soloman built for the LORD." In Leviticus 19:35–37, the injunction about weights and measures is even more specific than in Deuteronomy 25:13–16: "Ye shall do no unrighteousness in judgment, in meteyard, in weight, or in measure. Just balances, just weights, a just ephah, and a just hin shall ye have."

As Bruno Kisch notes, "In Greece and Rome (and probably also in ancient India) the original standard weights for comparison with the market weights were kept in the temples. The biblical expression 'shekel of the sanctuary' may hint at this fact also."[1] In the Roman Empire the standard weights were kept in special buildings called *ponderaria*, as they were later in the *Rentkammer* in Cologne. As early as A.D. 367, Pretextatus, prefect of Rome, ordered uniformity of weights throughout the empire. In A.D. 789 Charlemagne (just like King Otokar of Bohemia in the thirteenth century) ordered that "all should have equal and correct measurements and just and equal weights in the cities and in the monasteries."[2]

In England, Ethelred the Unready (968–1016), in dealing with "hateful illegalities to be earnestly shunned," put "false weights and wrongful measures" at the head of his list.[3] The Magna Carta (1215) contains King John's undertaking that "throughout the kingdom there shall be standard measures of wine, ale, and corn. Also there shall be a standard width of dyed cloth, russet, and haberject; namely [a width of] two ells with the selvedges. Weights [also] are to be standardized similarly." Though the intentions and the need were clear enough, this would prove to be a somewhat pious hope.

In "Select Tracts and Table Books Relating to English Weights and Measures," one can find a whole litany of complaints. For example, one Robert Thompson, in a report of about A.D. 1517, gives a complete "catalogue of abuses of false weights and measures which have been found out . . . in inns, markets, shops, warehouses and in many other places." His first item claims that he has

> found out six several sorts of weights that is to say: 1. Troy weight. 2. Avoirdupois weight. 3–4. Two sorts of Venice weights, one like the Troy weight, the other like the avoirdupois weights, but differing one quarter of a pound and less than a pound of true weight. 5. One other sort of weight called Ancell weight. 6. And one other sort of weights, which are made of stones, blocks of wood, and pieces of iron, made according to their own pleasure that use them and weigh with them.

Since at this point we are dealing with unrationalized measures rather than unrationalized language, I have transliterated Robert Thompson's report into relatively modern English.[4]

Thompson's second point speaks for itself. He reports that he has "seized and taken away many thousands of those weights, beside an infinite number of false measures of diverse and several kinds." The litany goes on and on. Item 11 lists by name nine different liquid measures that "are all contrary to the true standard." Item 15 deals with bakers whose bread is "too light by the fourth part." The next item deals with merchants involved in selling coal and wood, "whose sacks have wanted 3 inches in length and as much in breadth, besides the . . . ill filling of them." And finally, in item 23, he shows "that there have been many tradesmen that sell by yards and by ells which are made upon shop-boards [marked upon their counters], unsized [not assized], unsealed, and unlawful."[5]

Shopkeepers are not necessarily more honest today. Rather, the rationalization of weights and measures—together with the rationalizing of standards and packaging—has reduced the opportunities for dishonesty, at least in that direction. Some of us still remember with nostalgia the old-fashioned grocer weighing out his pats of butter. We should, however, also remember that such older practices encouraged the type of dishonesty of which Robert Thompson and many like him have complained for thousands of years.

When we consider how weights and measures began, it becomes easier to understand why, as with time measurement, they have proved so difficult to rationalize. The solar day, the lunar month, and the solar year seemed, at first, to be accurate criteria on which to base calendars and units of time. But for length, the units of measure were generally derived directly from man himself. Lengths based on the size of men would clearly be unequal. Some of the more primitive methods of weighing, such as judging the weight of goods by the load one could carry, now seem similarly uneven. Using a handful, or a pinch, or a spoonful were considered satisfactory enough for the early cookbooks that dealt alike with culinary or medical recipes.

There had long been a tool for measuring weight. Philip Rush claims that the beam scale, the first instrument used to compare the weights of two bodies, was initially used by the Sumerians about nine thousand years ago: "This instrument, called a *balance* in its most sensitive and accurate scientific form, is still the best of all weighing machines." In fact, it is "an invention of almost as much importance as the wheel."[6] The instrument is therefore not so much the problem as are the weights themselves, which cannot be approximated by any human measure. Rather, they must be checked against standard weights and used by honest merchants. Understandably, this did not always occur.

Our preliminary discussion will have demonstrated that early weights and measures were far too diverse to be reviewed with any profit in a single chapter. We will instead, like the writers of great epics, begin in the middle of things. We shall start with the eighteenth century, looking backwards at the main line of related developments. Colonial powers in general, and Britain in particular, were then attempting to rationalize weights and measures. We shall subsequently look at the French campaign to introduce metric reforms. Essentially, this reduces the scope of our investigation to a tale of two systems: the system of the British, who preferred to evolve, and that of the French, who preferred to impose a new and rational structure. After having earlier considered the French Republican calendar and the abject failure of its decimal system, both for the year and for time, the reader may be forgiven if he or she should suffer from a feeling of déjà vu. But the outcome on this occasion will prove to be very different.

THE BRITISH SYSTEM: A PROGRESSIVE RATIONALIZATION OF WEIGHTS AND MEASURES

During the Restoration and eighteenth century, the British leadership in advances in time measurement was naturally related to their worldwide interests in navigation and trade. Their involvement in rationalizing weights and measures was similarly motivated. Here, too, they preferred an evolution from existing standards, rather than a revolutionary approach like that of France, whose metric system was developed at the close of the eighteenth century. Before the French metric system, the measurements of length had been based on man himself, and they had differed widely from person to person and place to place. Moreover, many little kings, dukes, and bishops felt that they had the right to maintain their own standards of weight and measure, for obvious reasons of personal profit and pride.

The very words by which they have been denoted demonstrate that linear measurements were almost entirely anthropomorphic. The nail or the digit was a measure of approximately three-quarters of an inch and derived from the distance across the base of the nail of the middle finger. The inch (which means the *unce* or twelfth part of a foot) was generally a thumb's breadth. The palm (or four fingers) was approximately three inches, the hand four inches, and the foot twelve inches. The cubit (from the Latin for elbow) stretched from the elbow to the tip of the middle finger, and was one and a half feet; the yard (related to man's girdle or girth) stretched from the nose to the middle finger, and was three feet; the fathom (derived from a man's outstretched arms) was the distance between one extended middle finger and the other, and was six feet.[7]

While the ratios between lengths were fairly stable, the lengths themselves varied so much that they could hardly be considered standards in our modern sense. Lengths were either derived from specific people or were based on the average for several people. For example, Henry I of England (A.D. 1100–35) fixed the yard as the distance between his nose and the thumb of his outstretched arm.[8] King David I of

Figure 3. A British parody of Jakob Köbel's *Geometrei* (1575), which prescribed how the German rood of sixteen feet should be ascertained. (R. Vieweg, "Aus der Geschichte des Eichwesens bis zur ersten deutschen 'Vollversammlung'," *PTB-Mitteilungen* 79, no. 5 [1969])

Scotland (ca. 1150 A.D.) averaged the thumbs of a small, a medium, and a large man to arrive at the Scottish inch (figure 3).[9] But, of course, there was no correspondence between either the two kings or their standards of measurement. Indeed, right up to the Imperial Weights and Measures Act of 1824, the Scottish inch exceeded the length of the English inch. Furthermore, the Scottish mile of 5,952 (English) feet and the Irish mile of 6,720 (English) feet both exceeded the English mile of 5,280 feet.[10]

Feet, though they almost always contained 12 inches, were rarely of the same length. As late as the period between 1800 and the time when they adopted the metric system, the foot, in the townships of what is now Germany, ranged all the way from the English equivalent of 9.84 inches in Darmstadt to 14.01 inches in Frankfurt.[11] Since all the related measures varied proportionately, we can see that the British by comparison were operating a relatively standardized system. The French situation was probably even worse than that in the German states. G. G. Coulton, in *The Medieval Village*, writes: "In France, under the Ancien Régime, there were literally hundreds of variations in the legal measures; and, even in England, where our fourteenth-century kings tried to enforce a royal standard for the most important measures, exceptions had to be made for these manorial variations, and the peasant was left still at the mercy of local custom."[12]

The "manorial variations" continued long after Edward I, whose Act of 1305 was the first by an English king that set out tables of length and area measurement. Edward I had prescribed that "three grains of barley, dry and round, make an inch, twelve inches make a foot, three feet make an Ulna [or yard], five and a half Ulnae make a rod, and forty rods in length and four in breadth make an acre."[13] A manuscript of 1682, dedicated to James, duke of York, indicates for how long after the Act of 1305 the peasant was indeed left at the mercy of local custom when land was measured by the seemingly traditional rod: "A Pearch, or a Rod, or a Pole (by statut) must be 5 yards and an half; or sixteen feete and an half. But in some places of England they measure wth a pearch of 12 foote called Tenant right or Court measure. In other places they measure wth a pearch of 18, 20, or 24 foote, called Woodland Measure."[14] Lengths also varied according to the trade by which they were used. As late as 1751, Ephraim Chambers's *Cyclopaedia* defines *fathom* as follows: "There are three kinds of fathoms. . . . The first, which is that of men of war, contains six feet; the middling, or that of merchant ships, five feet and a half; and the small fathom, used in fluyets, fly-boats, and other fishing vessels, only five feet." Similarly, the *Oxford English Dictionary* still lists a variety of measures under *mile* and *ton*, a problem that the metric system avoids.

Though land measurement was particularly important when some 95 percent of the population was tied to the land, the standards for measuring area remained very uneven. The acre, which was originally the amount of land that could be ploughed in one day, had developed considerable variations. Edward I's standard clearly takes the source of the acre into account when he defines it as 4 rods in width by 40 rods in length. The width of 22 yards, or 4 rods, is the chain—a measure of length that is still used for the cricket pitch and by surveyors—while the length of 220 yards, or 40 rods, is the furlong (or furrow-long) of which there are still 8 in a mile of 1,760 yards. The most widespread measurement in the German states that compares with the acre (the day's work, or *Tagwerk*) was the *Morgen* (or morning's ploughing). As late as the period from 1800 to their adoption of the metric system, the *Morgen* varied all the way from the English equivalent of .50 acres in Frankfurt to 2.38 acres in Hamburg.[15]

Although the Roman measurements had also varied from place to place, the ratios between them have remained remarkably consistent right down to our own time. The Romans—like the British up to the eighteenth century and beyond—had 4 digits to a palm, 4 palms to a foot, 5 feet to a pace (which we would call a double pace), 125 paces to a stadium, and 8 stadia to a mile (derived from *mille*) of 1,000 paces or 5,000 feet. In order to ensure that the furlong of 220 yards could retain the ratio of one-eighth of a mile, Queen Elizabeth I increased the mile from 5,000 feet to 1,760 yards or 5,280 feet. The Roman digit (which becomes the British nail) was on average the English equivalent of .73 inches. This meant that the Roman foot of 16 digits averaged 11.62 inches. The eighth of a mile, as with the British furlong, was their stadium, the standard course for foot and chariot races. The Roman cubit also equaled 24 digits, and the Roman foot was either 16 digits or 12 *unciae* or inches. (In a long career the cubit has varied all the way from a Roman cubit of 17.496 inches to a Palestinian cubit of 25.25 inches.)[16]

When we come to the standards of weight, it is important to remember that the words *inch*, *ounce*, and *uncia* all have the same root meaning of one-twelfth. The nail, or digit (which was one-sixteenth of a foot) came to mean, in a comparable manner, a fraction of one-sixteenth in its own right. Thus, though the nail of a foot was three-quarters of an inch, the nail-of-a-yard was two and a quarter inches, or half of a finger that was four and a half inches. The nail-of-a-bushel was the sixteenth part of a bushel, the volume measure, and the nail-of-a-hundredweight was the sixteenth part of a hundredweight.[17] In fact the old term for a seven-pound weight—which was half a stone, a quarter of a quarter, an eighth of a half hundredweight, and a sixteenth of a hundredweight of 112 pounds—was a clove, from the Latin *clovus* or nail.[18]

As with measures of length, the ratios between standard measures of weight remain far more constant and predictable than do the values. The larger weights that we have been considering belong to the system of "avoir du pois," meaning "heavy goods," which is first mentioned in a statute of Edward III of 1340. Under the date of 1474, the *Coventry Leet Book; or, Mayor's Register* specifies the instruction of the king and his lords spiritual and temporal to the effect that the weight of 32 grains of wheat taken out of the midst of the ear makes a sterling penny, and twenty sterling pennies make an ounce avoirdupois, and sixteen ounces make a pound.[19] After many changes, including those of Henry

VII and Elizabeth I, the avoirdupois pound was declared by the Act of 1824 to weigh 7,000 grains divided into sixteen ounces of 437.5 grains each.

It would, of course, have been too simple to use the avoirdupois pound for all weights. Bullion and currency, in particular, had a system of their own. King Offa of Mercia (A.D. 757–96) had long ago instituted the moneyer's pound, which was called the tower pound after the Norman Conquest. This system was based on half of an Arabic coin, the silver *dirhem* of 45 grains. It provided a silver penny of 22½ grains, which set the standard of silver coinage in England until 1344 and of the weight system until 1527. The English silver penny was called esterling (and eventually sterling) because of its Eastern origin. The system had twenty pennyweights to the ounce of 450 grains and twelve such ounces to the pound of 5,400 grains. Since the word *pound* comes from the Roman *libra pondus*, or pound by weight, we can see that the moneyer's pound, from which our present pound derives, was originally a silver pound by weight. Pennyweight silver coins did in fact often do duty as weights.

The troy pound, named after Troyes in France, first appeared in English statutes in 1414. Over the next century it slowly took over from the tower pound, and there was a complicated system of conversion between the two. When Henry VIII finally abolished the tower pound in 1527, he ordered the mint to use only the troy pound for currency purposes. By about 1550, in the middle of the Tudor period, there had also arisen a confusion of merchants' pounds. As Rush and O'Keefe put it, "There was a pound of 6,750 grains, one of 7,680 grains, one of 6,992 grains, and in addition there was a merchant's pound of the Hanseatic league of 7,200 grains which merchants began to confuse with the 6,750-grain pound." All this, of course, was in addition to the troy pound of twelve ounces established in 1497 for "sylver, golde and breade," as well as spices and apothecaries' goods.[20]

Queen Elizabeth's reforms of 1588 provided standards that lasted until 1824. Her avoirdupois pound of 7,000 grains was divided into sixteen ounces of 437.5 grains each. The grains avoirdupois were of the same value as the grains troy, and the troy pound was set at 5,760 grains divided into twelve ounces of 480 grains each. Thus, although the avoirdupois pound was, and remains, considerably heavier than the troy pound, the troy ounce is heavier than the avoirdupois ounce. Elizabeth's solution had been to make the avoirdupois (wool) pound serve for everything except coin, bullion, and apothecaries' goods, for which the troy system was employed.[21]

Clearly even Elizabeth I could not eliminate entirely the use of different systems of weight for different commodities. Much the same was true insofar as volume measures were concerned. Despite the attempts over many years to rationalize the system, there were still at least four standard gallons in the England of 1824: the Winchester corn gallon of 268.8 cubic inches, Queen Anne's wine gallon of 231 cubic inches, an ale gallon of 282 cubic inches, and an older wine gallon of about 224 cubic inches. In accordance with their usual method of tinkering with existing standards rather than setting up a completely new system, the British government's Act of 1824 defined the gallon as containing ten imperial standard pounds of distilled water with the air at sixty-two degrees Fahrenheit and the barometer at thirty inches.[22] This was not followed, however, by the Americans, whose gallon is still based on Queen Anne's considerably smaller wine gallon of 1707.[23]

THE FRENCH SYSTEM: A TOTAL RATIONALIZATION OF WEIGHTS AND MEASURES

The British system of making ongoing adjustments to existing standards of weights and measures worked well enough while Britain controlled a large part of the world's land, trade, technology, and science. The weakness in such an evolving system had, however, long been evident. At the risk of making a dangerous generalization, one could suggest that the British system might be expected from the country of Sir Francis Bacon, whose experimental science evolved from experience rather than from authority or conjecture. The country of Descartes, on the other hand, might be expected to employ a more methodical and theoretical approach for arriving at a new system. Certainly that is what occurred. As early as 1670, Gabriel Mouton, vicar of Saint Paul in Lyons, had suggested a comprehensive decimal measurement system based on the length of one minute of arc (one-sixtieth of one degree) of a great circle of the earth. In the following year the French astronomer Jean Picard proposed that the unit of length should be based on the length of a pendulum beating seconds. It will be noted that both of these suggestions are radically different from a system founded on the length of the limbs of a man and a grain of barley, or the weight of a grain of wheat.

It can certainly be argued that the need for change on the European continent was far greater than in England. As late as the 1700s the European measures differed not only from country to country but even from town to town and from trade to trade. An official report of the British government published in 1864 states that "amongst the many well-founded grievances of the French people . . . which led to the French revolution . . . were the defective state and want of uniformity of their weights and measures [which], though nominally the same, varied in each province, and frequently also in many localities of the same province."[24]

Though the need and the concepts for the metric system certainly existed throughout the eighteenth century, the ancien régime clearly had neither the zeal nor the will to act. As early in the Revolution as May 8, 1790, however, the National Assembly of France enacted a decree to "deduce an invariable standard for all of the measures and all weights." Though they called upon the French Academy of Sciences together with the Royal Society of London to undertake this commission, only the French responded.[25]

In April of 1790 Talleyrand had already put before the assembly a plan for reform based on the length of a pendulum beating seconds at forty-five degrees latitude. When the provisional standard was adopted by a decree of 1795, however, the basic unit was equal to one ten-millionth of a quadrant of the earth's meridian. This unit was later called a meter from the Greek word *metron* meaning "a measure." Under the terms of the decree, Greek prefixes—such as *deca* and *hecto*—were given to the multiples of this unit, and Latin prefixes—such as *deci* and *centi*—were given to the subdivisions. From this single unit—which was decimally based for convenience and which in theory at least could be readily reproduced because of its natural origins—all standards of length, mass, and capacity were to be derived. Though complete accuracy was not achieved in developing the related standards, the gram was based on the weight of a cubic centimeter of water and the liter on a cubic decimeter of volume. In practice, even the meter did not conform exactly with its relationship to a quadrant of the earth.[26]

What has mattered most, of course, is the ease with which the decimal system can be used for money as well as for weights and measures, and the increasing uniformity that the metric system has brought to the modern world. But the path of those who endorsed metrification has been by no means easy. In Revolutionary France the attempt to apply the decimal system to time and the calendar fell away almost immediately. Even the metric system itself suffered severe early reversals. Though it had been made mandatory in 1795, its use was not enforced. Worse still, in 1812 Napoleon Bonaparte issued a decree that allowed the old units for weights and measures to return. In effect, the old unit names were brought back and the decimal ratios between units and their parts no longer held good. It only remained mandatory to define the measures being used by their metric equivalents. The resulting confusion could not, however, be allowed to continue indefinitely. On January 1, 1840, all systems other than metric were prohibited in France, and have remained so to this day.

Though they retained control of the metric system within their own country, the French have proceeded from the beginning like active evangelists who know that they have a good product to sell. And the increase in market share for their product has certainly been impressive. By 1880 seventeen nations, including the major powers on the European continent, had accepted the metric system, and by 1900 eighteen more nations had been added to this list. At the beginning of our own century, the two most important nations holding out against the public introduction of the metric system were Great Britain and the United States of America. However, since 1893 the U.S. prototype meter and kilogram have been considered the nation's "fundamental standards of length and mass,"[27] and by 1897 legislation had been enacted in Great Britain that permitted full use of the metric system.

The current status of the French and British systems of weights and measures would seem to be heavily weighted in the direction of the metric system. Within our own recent memory Britain, much reduced in international stature, has thrown in the towel, and the United States stands practically alone in employing the British system. One is reminded that by the middle of the eighteenth century, the British themselves—when they were the world's leaders in astronomy and the production of clocks—had held out virtually alone for no less than 170 years against the rationalization of the Gregorian calendar. They had, of course, long lost the battle without knowing it. A similar period of time will soon have elapsed since John Quincy Adams, as America's secretary of state, made an official prediction to his government that "the metre [sic] will surround the globe in use . . . and one language of weights and measures will be spoken from the equator to the poles."[28] When one compares the British stand on the Gregorian calendar with the present American stand on the metric system, one may well wonder whether history is not repeating itself.

Insofar as weights and measures are concerned, one hardly needs to outline the extent of the reaction against rationalization. In our own time, the rearguard action against metrification has been fought long and hard in Britain, Canada, and the United States. In Britain rumblings still exist, and in Canada there have been conciliatory noises about continuing to use both the gallon and the liter for gas. Nevertheless, metrification seems to be assured in both countries. One cannot help but wonder whether and for how long the United States might hold out. What is noteworthy is that—in contradistinction to the battle over the seven-day week and Bishop Ussher's chronology—there has been little if any recent interference from religious bodies with respect to rationalizing weights and measures.

RATIONALIZED MEASURES:
MAN NO LONGER THE MEASURE OF ALL THINGS

Measurements of length, as we have observed, originally varied considerably from place to place, and so did the equally important measurements of area to which they were related. We have also noted that the nail or digit, the inch, the palm, the hand, the foot, the cubit, the yard, and the fathom all derive directly from the measurements of man's body. His pace, in addition, is used as the basis for all the longer linear measurements. From this we can see that the ancient contention of Protagoras that "man is the measure of all things" may have a much more concrete foundation than we might have assumed. We may also anticipate from the *digit* that man's hand has provided the foundation for the measurement by numbers. But that abstraction does not contain the implication of considerable variables inherent in using man as the basis for all linear measurement.

Protagoras (ca. 481–411 B.C.) was the pupil of Democritus, who gave us the atomic theory. We are told by Diogenes Laërtius that Protagoras "was the first person who asserted that in every question there were two sides to the argument exactly opposite to one another. And he used to employ them in his arguments, being the first person who did so. But he began something in this manner: 'Man is the measure of all things: of those things which exist as he is; and of those things which do not exist as he is not.' "[29] As Socrates puts it in Plato's *Theaetetus*: "Come then let us question Protagoras, or someone else who holds the same opinions with him. Man, as you say, Protagoras, is the measure of all things, white, heavy, light, and every thing of that kind: for, as he contains the criterion of them within himself, in thinking they are such as he feels them to be, he thinks what is true to himself."[30]

Socrates belabors the point again in Plato's *Cratylus*, where he asks: If Protagoras insisted that "man is the measure of all things," would they be to me as they appear to me and to you as they appear to you, or do some things "possess a certain stability of existence?"[31] This is, of course, an important and unresolved question. It comes down to us from the eighteenth-century philosophy of Bishop George Berkeley (1685–1753) and the German Transcendentalists like Johann Gottlieb Fichte (1762–1814). Samuel Johnson, using an argument that Socrates might have envied, once said to a follower of Berkeley: "Pray, Sir, don't leave us; for we may perhaps forget to think of you, and then you will cease to exist."[32]

Yet Johnson argues elsewhere (like Protagoras, of the three material dimensions) that time at least is in the mind of the beholder: "Time is, of all modes of existence, most obsequious to the imagination; a lapse of years is as easily conceived as a passage of hours."[33] And in a letter to Boswell of September 19, 1777, Johnson writes: "Depend upon it, Sir, when a man knows he is to be hanged in a fortnight, it concentrates his mind wonderfully."[34] Johnson—like many in his century and since—may be reacting against the precise measure of clock time.

When Horace (65–8 B.C.), in his "Epistle to Maecenas," deals with the point made by Protagoras, he is clearly cognizant that, on one level at least, the reference may be to the actual linear measures that are based on man. Horace concludes Epistle 1.7 with the statement: "It is meet that every man should measure himself by his own rule and foot."[35] It would therefore seem from Horace's version, and probably also from that of Protagoras, that our proverbial saying "The measure of mankind is man" contains implicitly an inherent sense of the variability of linear and other measures.

Our own standards of temporal and linear measurement have recently been separated not only from man but also from our earth. We have previously noted that in 1967 the second itself was finally divorced from the rotation of the earth and redefined in terms of the vibration of the cesium 133 atom. Seven years earlier, the meter had been similarly removed from its relationship with the measurement of the earth and redefined in terms of the wave length of the element krypton. Similarly, William D. Johnstone notes that "the inch has been corrected to equal 41,929.399 wavelengths of krypton 86, measured at 760 millimeters pressure and 15 degrees centigrade."[36]

I have suggested that although linear measurements and numbers derive directly from the human body, time does not. Since this may be wrong in one important particular, it is worth considering the relevant facts. According to the medieval measures of time in the table of Papias in Du Cange, as it appears under *atom* in the *Oxford English Dictionary*:

47 atoms of time	= 1 ounce	= 7½ seconds	(modern)
8 ounces	= 1 ostent	= 1 minute	"
1½ ostents	= 1 moment	= 1½ minutes	"
2⅔ moments	= 1 part	= 4 minutes	"
1½ parts(4 moments)	= 1 minute	= 6 minutes	"
2 minutes	= 1 point	= 12 minutes	"
5 points	= 1 hour	= 1 hour	"

What should strike us about the terms used is that four of them at least suggest very short periods of time, even though we might now think of them as denoting longer periods in relation to modern technology. They are, in descending order of temporal length, *point, minute, moment,* and *atom*. Clearly each in its turn has represented a minute period of time. The point speaks for itself, and the minute when differently pronounced indicates its derivation. When Alexander Pope speaks of the atomic philosophers in the "Argument" to *Dunciad*, book 4, he calls them "Minute Philosophers." For him, the terms would seem to have been interchangeable.

As we descend in the order of minuteness, the term *moment* is still with us as the reference to a point in time. With the *moment* and the *atom*, we arrive at temporal terms that may well derive from movements of the human body, and indeed relate space with time. The term *moment* does, in fact, derive from *movement* and refers to a physical movement of the shortest duration. Hence, a quotation from 1340, in the *Oxford English Dictionary*, defines *moment* as what is considered the shortest possible human movement for measuring time: "A moment of tyme es nan othir thyng, Bot a short space als of a eghe twynkling."

The *Oxford English Dictionary* gives two meanings for *atom*, from the Greek *atomos*: "the twinkling of an eye and that which cannot be divided." The *OED* notes that the Latin *atomus* also has the meaning of "twinkling of an eye," and that this meaning already existed in the Greek *atomos*. The *OED* also gives a translation from the Greek *atomos* in 1 Corinthians 15:22 (actually 15:52) which reads: "in a moment, in the twinkling of an eye." As Trevisa puts it, "Dyuydynge [dividing] . . . of tyme passyth no ferder than Athomus."

At this "point in time," it must remain a moot point whether the blinking of an eye provided the first temporal concept for a physical particle so small that it could not be further divided. The root meaning of *atomos* suggests something indivisible. It is therefore also possible that the tiniest physical particle suggested the image through which to describe a temporal period so short as the twinkling of an eye. Temporal and spatial images are readily enough exchanged. We may note again that the spatial image of progress—as in a kingly progress, or "going a progress through the guts of a beggar"—did not change into the modern temporal image of a progress through time until the seventeenth century. When we consider "space-time," we shall find that the relationship between the three dimensions of space and the fourth dimension of time is very much a part of our current century, and our model of an expanding Einsteinian universe.

The original measure of many if not all things was certainly man. We have seen how he provided the measure of spatial units, as well as the numerical digits at which we will be looking in the next chapter. We may also suspect that he provided—through the shortest of moments (or movements), such as the "twinkling of an eye"—the earliest basic unit for time. This human-based measurement for the shortest period of time still remains remarkably universal insofar as language is concerned. We need go no further than German, French, and Italian, as well as English, to see that *Augenblick, en un clin d'oeil,* and *in un batter d'occhio* mean equally "in the twinkling of an eye" or "in a moment (or movement)." Martin P. Nilsson, in *Primitive Time Reckoning*, gives as the expression for the shortest period of time among Malays, Javanese, and Achanese: "a blink of the eyes."[37] We may assume therefore that man really has been used as the basis for temporal as well as linear measure. Rationalization has, however, now been taken so far that man's original relationship remains only in the etymology of the words through which such measures continue to be described.

V The Ascendancy of Hindu-Arabic Numerals

In the early systems of numbering there was no abstraction of the numbers themselves. Among the Fiji Islanders, as Karl Menninger explains, ten coconuts was *koro* and one thousand coconuts was *saloro*, but ten boats was *bola*. A slightly more abstract system developed in British Columbia, where a group of native Indians have different number terms for animate things, round things, long things, and days. For example, three animate things are *yatuk*, three round things are *yutsqsem*, three long things are *yututs'ak*, and three days are *yutqp'enequls*. Thus three men are called "*yatuk* men," but three days are called "*yutqp'enequls* days."[1]

Although one would be right to think that the English language has developed far in rationalizing the abstraction of numbers, vestiges of an older system also remain in our language. We refer to a school of fish, a pride of lions, a pod of whales, a stand of trees, a gaggle of geese, a pack of wolves, a herd of deer, or a flock of sheep. We know instinctively that we should not speak of a pod of lions or a flock of fish. And there are also vestiges of older terms related to the number two. We refer to a brace of pheasants, a yoke of oxen, a couple of hounds, or a pair of shoes, but not to a brace of shoes or a pair of pheasants. The word *team* probably derives from the Old High German *zoum*, the number word for two or more beasts of burden harnessed together. We can talk of a team of dogs or football players, but not of a team of wolves or canaries. We speak of a trinity or quaternity of gods, but we use triplets or quadruplets for people. Similarly, a duet, a quartet, and a sextet are normally reserved for voices, instruments, or musicians.

In addition to there being number words related to specific types of animals or articles, concrete representations of number were similarly used in early written records. Thus the Sumerians, as early as the eighth millennium B.C., recorded their possessions with small tokens of baked clay. They used, for example, lenticular discs for sheep and cone-shaped tokens for small measures of grain. Since a sphere represented ten measures of grain, three cones and a sphere would represent thirteen measures. With the rise of urban culture, the Sumerians made complex tokens, such as small models of tools and vessels. These were used in a numbering scheme that, at first, involved no abstractions. As Bertrand Russell put it, "It must have taken ages to discover that a brace of pheasants and a couple of days were both instances of the number two." According to archaeologist Denise Schmandt-Besserat of the University of Texas, it took the Sumerians some five thousand years after 8000 B.C. to achieve this most important aspect of rationalization.[2]

The philosophy of Plato might suggest that the idea of the number two would have existed in its abstract form—much like the idea of a bed and a table in book 10 of his *Republic*—before the concrete forms came into being on this earth. Such propositions have, however, been sorely tested by the introduction of modern chronologies. According to Schmandt-Besserat, the Sumerians did not dream up numerals until 3100 B.C., shortly after they began writing on tablets of clay. Then they used a small wedge for 1, a small circle for 10, a large wedge for 60, and a large circle for 360. There were as yet no specific symbols for intermediary numbers like 2, 3, or 4. The all-important rationalization making abstraction possible had, however, been achieved, and the Greeks would later work out much of the rest.[3]

Georges Ifrah explains how armies in Ethiopia used to calculate their casualties by using a pile of stones: "Before going off on an expedition, each warrior placed a stone on a pile; afterward, each survivor took a stone off the pile, and losses were judged by the number of stones remaining."[4] Menninger explains how early traders would convey the idea of a number to primitive people who possessed as few as three number words. To convey the idea of twenty buffaloes one might count out twenty nuts from a bag; to exchange twenty sheep for forty twists of tobacco one might have to make twenty individual exchanges of two twists for each sheep. But such ideas are by no means entirely primitive. In Spain, at one time, the innkeeper would toss a pebble (*chinas*) into the hood of a customer's cloak for each drink he served. Thus *echar chinas* (to throw pebbles) still means "to chalk something up" to a customer's account. The related English terminology reminds us how widespread has been the use of the tally and the chalkboard in comparable circumstances.[5]

Robinson Crusoe kept a faithful tally of the twenty-eight years he spent on his island in the Caribbean, "in the latitude of 9 degrees 22 minutes north of the line." Since he is particularly anxious not to "forget the Sabbath days," he carves on a post: "I came on shore here the 30th of September 1659." He adds that on the sides of the post, "I cut every day a notch with my knife, and every seventh notch was as long again as the rest, and every first day of the month as long again as that long one; and thus I kept my calendar, or weekly, monthly, and yearly reckoning of time."[6]

EARLY NUMBER SYSTEMS

The digits of the hands are an obvious basis for number words. Primitive people, however, also used other parts of the body, and the early meanings are often still discernible in their words for the related numbers. For example, Ifrah quotes from the Bugilai in New Guinea, whose word for one means left hand, little finger; two, next finger; three, middle finger; four, forefinger; five, thumb; six, wrist; seven, elbow; eight, shoulder; nine, left breast; ten, right breast.[7] It is quite clear that people learned to count on their fingers, and children of many cultures still do. But this use of the "digits" should not be confused with the more complicated systems of finger counting that were developed by most advanced civilizations of the past. In such systems, numbers as high as one hundred thousand can be counted on the hands.

Georges Ifrah describes a system used in China as late as the nineteenth century, in which numbers up to one hundred thousand could be counted on one hand: "Each joint is divided into three parts: left, right, and center. Each finger is then associated with the nine units of a decimal order, corresponding to the nine divisions of its joints: the little finger with the first nine units, the third finger with tens, the middle finger with hundreds, and so on."[8] Karl Menninger reports that the "Romans had a method of representing the numbers from 1 to 10,000 on the fingers of both hands—a form of 'finger-writing.'"[9] Ultimately, the abacus and Hindu-Arabic numbers would eliminate finger-writing, though it still remains in a modified form among Arab and Indian traders in the Middle East.

Since the practice of finger counting appears to have been so widespread, even among "illiterate" people, few textbooks seem to have been written on this subject. A noteworthy exception was produced by the Benedictine monk the Venerable Bede (d. 735 A.D.). His "On Calculating and Speaking with the Fingers" forms the introduction to the important work on chronology *De temporum ratione*. A central purpose of Bede's book was to calculate the variable dates for Easter Sunday. The Venerable Bede begins his work: "Before we begin, with God's help, to speak of chronology and its calculation, we deem it necessary first briefly to show the very necessary and ready skill in finger counting." He then proceeds:

1: If you wish to say "one," you must bend the little finger of your left hand and place its tip on the palm;
2: for "two" lay down the ring finger next to it.

He continues through to 9,999 with the use of both hands. Beyond that, one proceeded by added gestures—such as "60,000: grasp the left thigh from above"—to 1 million.[10]

Finger counting was a very widespread language—admittedly with many "dialects"—which seems to have died out almost without a trace. One rarely if ever hears of anthropologists—who are usually so vocal concerning the fate of native dialects—bemoaning its demise. Yet finger counting obviously gave a high level of "numeracy" to many people and went quite beyond the capabilities of such awkward written numbering systems as the Roman numerals or the alphabetic system of the Hebrews. Ifrah devotes a chapter to "The First Calculating Machine: The Hand." Among his illustrations are two Roman tokens from the first century A.D. One token shows the number nine represented by a finger sign on one side and the Roman numeral VIIII on the other; the other token shows a man representing the number fifteen on one hand, using the same system of finger signs.[11] The decimal system clearly grows out of our having learned to count on our ten fingers. But finger counting developed a long way beyond these simple roots, as they were described in *Fasti III* of Ovid (43 B.C.–A.D. 18): "A year was counted when the moon had returned to the full for the tenth time: that number was then in great honour, whether because that is the number of the fingers by which we are wont to count, or because a woman brings forth in twice five months, or because the numerals increase up to ten, and from that we start a fresh round."[12]

Georges Ifrah notes that Moslems use the finger joints and balls of the thumbs to enumerate the ninety-nine (three times thirty-three) attributes of Allah: "Moslems use this method when a rosary is not available. It is very old, and probably preceded the use of the rosary [three rows of thirty-three]. In several traditional stories the Prophet proscribes the use of beads or pebbles for reciting litanies and recommends counting prayers or praises of God on the fingers."[13]

In discussing the widespread use of tally sticks for denoting numbers, Ifrah and Menninger show that the Roman numerals may well derive from the notches that originally appeared on tally sticks. For instance, *X* for ten derives from the crossed *I*, and *V* for five from *X* cut in half. In Egypt, it was a common practice to denote the fraction for a half by an article, like a measuring vessel, divided horizontally through the middle.[14]

Early symbols for large numbers—usually derived from prolific living things—often became hieroglyphs for a particular number. For example, the Egyptian symbol for one hundred thousand is a tadpole, and its name, *hfn*, also means "innumerable." In China, *wan*, the scorpion, is both the symbol and the number for ten thousand. In Greek, *myrios*, countless, became *myrioi*, the number word for ten thousand. It probably derives from *myrmex*, the ant. The Indian name for the lac louse, which is found in "countless" numbers, seems to derive from the Sanskrit number word *laksa*, which is a lakh or one hundred thousand.[15] This number was evidently intended to suggest "countless." By the time that I had arrived in India during World War II, however, bourgeois types and their potential brides were categorized by the numbers of lakhs of rupees that they were worth.

Before the introduction of the Hindu-Arabic numerals, systems like the Roman, the Hebrew, and the Minoan were employed for writing down large numbers. Since they were hardly suited to being used for addition or subtraction, not to mention multiplication or division, Ifrah shows how calculations would first be carried out in finger writing.[16] The Hindu-Arabic numbers—because they could also be used for calculations—had more to offer than simply being a decimal place-value system. The Minoan counting system was already using a decimal place-value system in the second millennium B.C. It is also readily understandable, as the following denotation for 3,496 will show:

One can hardly envisage, however, undertaking mathematical calculations with such a numbering system.[17]

INDIAN AND ARABIC INFLUENCES

In dealing with eschatologies, we have already noted that the Indian eternalistic eschatologies embraced numbers of a completely different order of magnitude from virtually any other. Each *kalpa*, for example, involved a mythical period of 4,320 million years. It is also clear that the numbers were intended to impress by their patently suprahuman qualities. We learn, for example, that at the "manifestation" of Gautama as the Buddha "the Queen Maya Devi, will be attended by ten million ladies-in-waiting. Hundreds of thousands of holy men and hundreds of thousands of millions of the enlightened will pay homage to the Buddha. His throne is constituted of good works performed during hundreds of thousands of millions of *kalpas*. But the great lotus which blooms during the night of the Buddha's conception opens its blossom to an extent of 68 million miles."[18]

Such great numbers cannot be treated any more seriously than the constricted chronology of Bishop Ussher. But the Indian eternalistic eschatology clearly demanded a system and an expertise capable of handling such figures. When Buddha courted Gopa, her father demanded that he first pit himself against the great mathematician Arjuna. In his turn, Arjuna demanded that Buddha list all the numerical ranks above 100 *kotis*. *Koti* was the name of the seventh rank, meaning

10^7 or 10 million. At an early date, India already had names for the ranks leading as high as *laksa* (10^5), 100,000 and *prayuta* (10^6), 1 million.[19]

The essential difference between the Indian numerals and an early place-value system like the Minoan is that each Indian figure up to nine has a discrete notation. When to this was added the zero and the decimal point, the Hindu-Arabic figures would eventually lead to the rationalization of the numbering systems throughout the world. Without this rationalization it would also be hard to envisage our modern advances in mathematics. The term *Arabic* in the Hindu-Arabic numbers derives from the fact that the Arabs were the traders responsible for the westward migration of the number system.

Menninger's illustrations show the appearance and morphology of the numbers as they moved westward (figure 4). He starts with the Brahmi numerals and proceeds through the Indian (Gvalior), the West Arabic (gubar), and East Arabic (still used in Turkey). Of these, the West Arabic numerals are clearly closest to our own. By the sixteenth century Albrecht Dürer was employing numbers derived from the Brahmi numerals that differ little from what we would use today. The 1, 2, and 3 of the Brahmi numerals, which Menninger illustrates, were written –, =, ≡. It is generally accepted that the figure 2 was formed by joining the two horizontal lines. In Germany, 2 often took the form Z when carved in stone. Similarly, the 3 was formed by connecting the three horizontal lines with curved ends.[20]

Georges Ifrah shows variants of the West Arabic "gobar numerals," as they appeared in books between the fourteenth and seventeenth centuries. He also displays the Codex Vigilianus, a Spanish manuscript dated A.D. 976, which contains the earliest known example of Hindu-Arabic numerals in Europe. At first, the Hindu-Arabic numerals seem to have spread across Europe not so much through manuscripts as through the improved counting board. This was said to have been introduced by the Frenchman Gerbert of Aurillac, who is known to have been in Spain during A.D. 967–70. Gerbert is regarded as having been the first great scholar to have spread the uses of Hindu-Arabic numerals and the astrolabe in Europe. He was elected pope in 999, as Sylvester II, and died in 1003. His counting board used counters called *apices*. Each had a value of between one and nine marked on them, usually in Hindu-Arabic numerals.[21] It was the empty space on the

THE ASCENDANCY OF HINDU-ARABIC NUMERALS

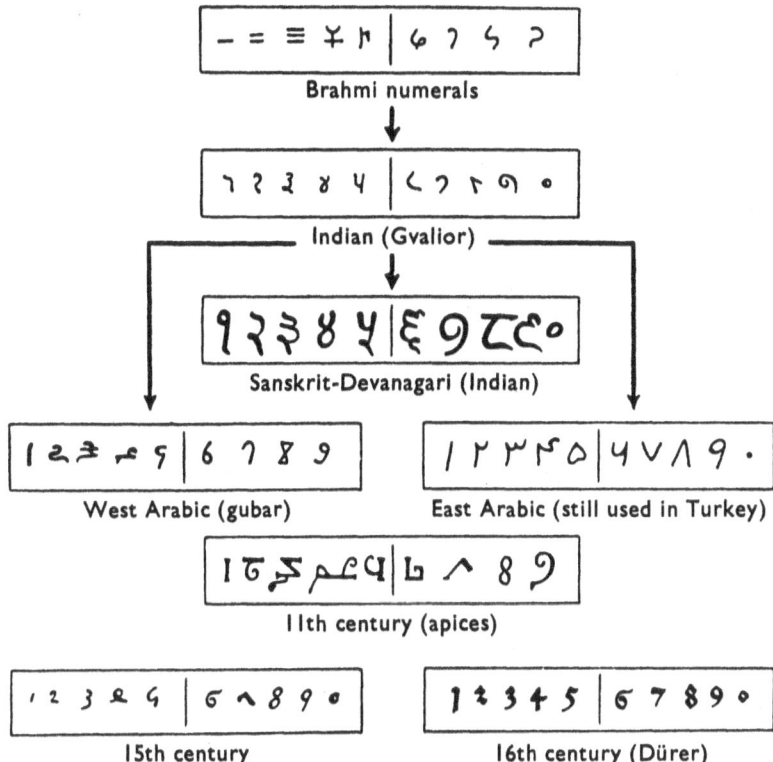

Figure 4. The Family Tree of the Indian Numerals. (Karl Menninger, *Number Words and Number Symbols*, trans. Paul Boneer [Boston: MIT Press, 1969])

counting board that gave us our zero. The Sanskrit *sunya* (empty), of the sixth to eighth centuries, became the Arabic *as-sifr* (the empty) in the ninth century, and the Latin *cifra* in the thirteenth century. From this we derive the German *Ziffer*; the Italian, French, and English *zero*; and the English *cipher*.[22]

The abacus is still used in parts of the Orient and the Middle East. It is a calculating device of ancient origin that has developed through many forms. In one form, still current, each upper bead represents five units and each lower bead represents one unit (figure 5). Only the beads pushed against the center strut are counted, and they are read off in the order in which the column stands. As the nine Hindu-Arabic figures and zero took over from the Roman numerals, the abacus as a means of calculating was gradually abandoned in the West. In England

Figure 5. In this Chinese abacus, the numbers in the upper frame are fives; those in the lower frame are ones. To make a number, move the beads to the central strut. The number shown in the illustration is 6,543,210. (Author's collection)

and Germany it was last in general use in the seventeenth century. At about that time, the systematic use of decimal fractions began in Europe with Simon Stevin's *Thiende* (1585), and in a less clumsy form in John Napier's second book on logarithms, *Mirifici logarithmorum canonis constructio* (1619). Decimal fractions were, however, used earlier by medieval Chinese mathematicians. By the early fifteenth century, the Persian astronomer al-Kashi had already computed pi to sixteen decimals.

ALTERNATIVES TO THE BASE TEN NUMERICAL SYSTEM

The fact that Hindu-Arabic numerals are now used virtually throughout the world should not lead us to believe that the rationalization of numbers to a base-ten system was either inevitable or readily accepted. We still have active vestiges of base-twelve, -twenty, and -sixty systems among the numbers with which we deal. The vigesimal or base-twenty system probably originates, like the decimal system, with the number of human digits, but including the toes. The Mayas counted by twenties and by powers of twenties. Their more elementary numbers are written in a manner that is quite simple. For the numbers one through nineteen, one through four are represented by up to four dots, and these are brought together in a group of five by a horizontal bar. For example, ⋮ equals six and ⋰ equals seven. Though one might expect that vigesimal number systems would predominate in warmer countries where toes were always in evidence, Menninger notes that "the 20-grouping is most strongly rooted in the north of Europe, in Iceland, Denmark, and England."[23] Expressions like "quatre-vingts," or the expectation that man might live for threescore years and ten, or fourscore at the utmost, strongly suggest an older system of numbers. It used to be customary to make a score or notch on a stick for every twenty when counting sheep or large herds of cattle, and the term may be related to this action. The term *score* was, however, also frequently used for a distance of twenty paces, or for a weight of twenty pounds of pigs or oxen, or of twenty to twenty-one pounds of coal.

In Charlemagne's coinage, twelve *denarii* of twenty *solidi* each were struck from one pound, or *livre*, of silver. This monetary system, which combines the twelve- and twenty-number groups, was not given up in France until the Revolution. The English system has reversed the numbers of shillings and pennies, but in other respects it has remained the same until very recently. The abbreviations for pounds, shillings, and pence derive from the related Latin words: *libra*, *solidus*, and *denarius*. Though the British fought long and hard to retain this system, it is now slowly passing into history, and the change to metric pounds represents yet another form of rationalization.

The twelve pence to the shilling reflects the widespread base-twelve number system. This system is still very much with us, though it is now retreating against the rising hegemony of decimal numbers. We can readily sense that the number twelve must have been of great significance. The months, the hours, the apostles, the tribes of Israel, the labors of Hercules, and the signs of the zodiac all bear witness to this. And so do counting by the dozen and the gross, just as counting by "ounces," which really means one-twelfth, and by pennies, which were one-twelfth of a shilling.

Karl Menninger suggests "another reason for the universality of the number 12 as a measure of quantity—namely the 'excess.'" There certainly was a widespread practice related to excess, particularly in Western Europe. Obvious examples—for the most part still with us—are the twenty-one pounds in a score of coal; the baker's, devil's, long, and printer's dozen of thirteen; the German *acht Tage* for a week; the French *quinze jours* for two weeks, the year and a day in legal terminology, and the thousand and one Arabian nights of Scheherezade. But this does not seem enough to convince one of Menninger's argument that "as a result . . . not ten, but twelve as 'ten with an excess' may have come into common use."[24]

On the subject of a measurement system based on twelve, Georges Ifrah seems more persuasive. He argues that the twelve-part numbering system, like the related sixty-part numbering system, goes back to a long-established and widely used finger-counting method. As he puts it, the "duodecimal finger-counting method used in India, Indochina, Pakistan, Afghanistan, Iran, Turkey, Iraq, Syria and Egypt" involves counting to twelve. One does this by using the thumb of the right hand, beginning with the outermost of the three bones at the tip of the little finger. The sexagesimal finger-counting method, still used in most of the same countries, is complementary to the duodecimal one. It uses the left hand to indicate twelve, twenty-four, thirty-six, forty-eight, and sixty by closing down each of the five fingers, starting with the little finger and finishing with the thumb.[25] Understandably, Ifrah feels that the duodecimal finger-counting method may have been a factor in leading the ancient Egyptians to divide day and night each into twelve unequal or temporal hours. It may also have "led the Sumerians, and the Assyrians and Babylonians after them, to divide the cycle of day and night into twelve equal parts (called *danna*, each equivalent to two of our hours), to adopt for the ecliptic and the circle a division into twelve *beru* (30° each), and to give the number 12, as well as its divisors and multiples, a preponderant place in their various measurements."[26]

As we have found before in this study, rationalization almost always results from the modification of existing systems. In the battle between the decimal and the duodecimal number systems, the decimal system seems to be winning out. One factor that should not be overlooked is that the decimal finger-counting system came to predominate in the West. The forces of history always play their part, but we should not make the mistake of assuming that they always produce the optimum results. The *Oxford English Dictionary*, under "Twelve," quotes Isaac Todhunter's text *Algebra* (1875): "The number ten has only two divisors . . . the number twelve has four. . . . On this account twelve would have been more convenient than ten as a radix."

The sexagesimal number system was widely used by the Babylonians. Their circle was divided into 360 degrees, possibly because the year is divided into approximately 360 days. Each degree was further divided by the sexagesimal number system into minutes and second minutes. From these divisions, we derive our modern minutes and seconds. Ptolemy (fl. 127–141 or 151 A.D.) used sexagesimal fractions in his great work on astronomy that entered our civilization through the Arabs. It has come down to us as the *Almagest*, the fundamental textbook of astronomy for about fourteen hundred years, until the time of Copernicus.

We rarely think in terms of the base-sixty numbers today. However, the word *Schock*—which is current in German but archaic in English—suggests that this has not always been the case. Cassell's *German and English Dictionary* defines the noun *Schock* as follows: "heap, shock; three-score; a (large) quantity; mass, lot; *zwei Schock*, six score." The verb *schocken* is defined as: "count by sixties; place in heaps of sixty." This is a term used both for specific numbers and also to indicate large but unspecified numbers. Our modern English equivalent would be: "I have done it scores of times," or "I have done it dozens of times." The related English word *shock* would also seem to be a vestige of the sexagesimal system. It is defined in the *Oxford English Dictionary* as "A lot of sixty pieces. (Used with reference to certain articles of merchandise originally imported from abroad.)" A quotation from 1674 says: "Many small wares called Habberdashery [sic] . . . are sold by Dozens, Scores, Shocks." Under the explanation for the derivation of the term stands: "shock of corn, group of 60 units."

The duodecimal and sexagesimal number systems have held out most strongly in the measurement of time: in our months, days, hours, minutes, and seconds. They also remain strongly entrenched in the related numbers for dividing the circle and the globe, which have come down to us from Babylonian astronomy. In these areas, the powers of history seem to be stronger than those of rationalization. But the Hindu-Arabic numerals now also have their own power of history behind them, as we have discovered with the recent introduction of computers. Though their numbering system is tied to a binary base, no computer manufacturer would dare to impose this on the public.

RATIONALIZING NUMBERS AND MATHEMATICS

Even more than the English language, the Hindu-Arabic numerals are now the lingua franca of the modern world. In a chapter entitled "Was

Writing Invented by Accountants?" Ifrah shows how writing seems to have been invented in the second half of the fourth millennium B.C. He argues that writing appeared at that time in both Mesopotamia and Elam for strictly economic and utilitarian purposes. It seems to have been developed by accountants because they were obliged to deal with economic operations that were both too complex and too varied to be entrusted to a single memory. An example was the great construction projects in the Sumerian city of Unruh.[27]

Sumerian tablets from 3200–3000 B.C. show numerals in conjunction with writing signs and cylinder-seal impressions denoting particular articles. Among these are animals and jars that clearly suggest a form of documentation related to accounting. As Ifrah puts it, the first use of such documents was primarily as a memory aid: "All known archaic Sumerian tablets [are] ... documents used by temples. Writing was probably invented and developed because at that time temples were the only economic authorities for all of Sumer, where constant, systematic production of large surpluses required increasingly complex and centralized redistribution."[28]

A second and almost equally important reason for being able to write down numbers is related to astrology. The symbiotic relationship between numbers and astrology is reflected in the fact that the Latin word *mathematicus* means both a mathematician and an astrologer. The first specific date that we have for the Mesopotamian records of eclipses translates into 747 B.C., and comes from the Greek astronomer Ptolemy. It is thought, however, that the Mesopotamians both named and recognized a number of constellations as early as 3000 B.C. Though astrology was used among the Greeks to forecast the future for individual human beings who paid for these services, the earlier "mundane astrology" of the Mesopotamians dealt with the prospects of kings and states.

The casting of a horoscope (Greek *horoscopos*, from *hora*, "time," and *scopos*, observer) is concerned with the relationship between the seven "errant" planets and the zodiac at the moment when a child is born. Thus the study of astrology leads directly into astronomy. The Greeks used spherical geometry to explicate the wheeling of the celestial bodies in their presumed circular orbits, and by doing so gave to astrology the stamp of Greek geometry rather than Mesopotamian arithmetic.

Astronomy came to Europe through the Arabic translation of

Ptolemy's *Almagest*. In this respect the fall of Toledo to the Christians in A.D. 1105 provided a catalyst for the scholars who came to Spain to translate the Arab works. In this they were helped by some of the ten thousand Jews who lived in Toledo. In the middle of the twelfth century, Gerard of Cremona came from Italy looking specifically for Ptolemy. By the time that he died, in 1187, he had translated not only the *Almagest* but also more than ninety other texts. In 1276 Alfonso the Wise ordered the preparation of the Alfonsine Tables to be updated from the star table of Ptolemy. These were to become standard astronomical reference works for the next three centuries and did much to spur further interest in astronomy.

Another of the translators of Arabic material was Adelard of Bath, who is thought to have been in Spain about A.D. 1110–25. His writings include works on the abacus and the astrolabe, as well as a translation of Maslama's edition of the astronomical tables of al-Kwarizmi (d. ca. 850). Adelard's small book entitled *Liber Algorismi de numero Indorum* (ca. 1120) is the first written work we have in the West dealing with the Hindu-Arabic numerical system. Al-Kwarismi's name has given us the term *algorithms*, and it is from the title of his book *Kitab al jabr wa'l muqabala* that we derive our term *algebra*.

Equations and methods for reducing them were first introduced in Babylonia. As early as the Hammurabi dynasty (1800–1600 B.C.), a number of cuneiform mathematical tables dealt with problems that are now classified as algebra. With the later Greek mathematics, however, algebra took on a geometric aspect, probably because of the geometric language related to Euclid's *Elements* (ca. 300 B.C.). After the translation of the Arabic texts, Cardano's *Ars magna* (1545) was the most important algebraic work of the Renaissance, and contained the solution of the cubic equation. Around 1515 Scipione del Ferro had succeeded in solving the cubic equation. He did not reveal his formula, but after his death it was passed to his pupils. (We shall refer again to this quality of secrecy when we deal with the guilds, whose trades were called *mysteries*, and with time and method study, the secrets of which were not disclosed until after factories came under the control of paid managers.)

François Vieta was the first to introduce an algebraic sign language, but the present usage of denoting unknowns by the last letters (x, y, z) and known quantities by the first letters (a, b, c) is attributed to Descartes. Though they had been used sporadically earlier, he was the first to make a practice of writing exponents in the modern form. The development of algebra was also stimulated further by the introduction of such other new fields in mathematics as the analytic geometry of Descartes and the calculus of Newton and Leibniz.

THE INFLUENCE OF MOVEMENT AND TIME ON MATHEMATICS

Another great spur to mathematics and many other related aspects of human endeavor came from the translation of Alhazen's *Optics* from the original Arabic. Alhazen, who had been born in Basra in A.D. 965, had drawn on the works of Aristotle, Galen, Euclid, and Ptolemy. He had, however, reversed the normally accepted principle that light goes from the eyes to the object that is being viewed. Thus Alhazen argued that as something moves closer to the eye, the cone of rays that enters the eye becomes larger and subtends a larger angle. This, and only this, makes the object viewed look larger as it moves closer to the eye of the beholder. From Alhazen's insight, we have the foundation for the concept of perspective, which, in Jacob Bronowski's words, "is the new idea which now revivifies mathematics."[29]

A translated manuscript of Alhazen's *Optics* in the Vatican Library has been annotated by Lorenzo Ghiberti (ca. 1378–1455). Ghiberti made the famous perspectives for the doors of the Baptistry in Florence, for which he won the competition in 1401. Filippo Brunelleschi (1377–1446), the first great architect of the Italian Renaissance, was probably also the first of the "Perspectivi." His design for the Baptistry doors was, however, placed second to that of Ghiberti. As Bronowski puts it, the new idea of perspective "was not simply to make the figures lifelike, but to create the sense of their movement in space." This, of course, gives a feeling for a fourth dimension or a movement through time. It changes painting radically from that period onward, for "above all, we feel that the painter's eye is on the move."[30]

The change in painting was remarkable. Bronowski shows a fresco of Florence—painted about 1350, and before the advent of the "Perspectivi"—in which "there is no attempt at perspective because the painter thought of himself as recording things, not as they look, but as they are." Bronowski contrasts his example from the fourteenth century with one by Albrecht Dürer, who traveled to Italy in 1506 to learn "the secret of perspective." As Bronowski puts it: "All the natural details in which Dürer delights are expressions of the dynamic of time."[31]

When Gotthold Ephraim Lessing wrote his *Laokoön* in 1766, he could already assume that, unlike poets, painters and sculptors had to make the best of the moment in time to which they were limited. Therefore, they should make the most of a work's potential dynamic by illustrating the moment just before the climax of an action. Examples are the fold in a gown just before the wearer is about to move, or the figures in *Laokoön* just before the snakes are about to make them cry out in agony.

Though Lessing's world was dominated by time, in Greece, Islam, and medieval Christianity the universe tended to be unchanging, static, and timeless. Since these civilizations thought in terms of perfect order, the most perfect shape for them was the circle. Eternity, like the harmony of the spheres, was illustrated by the circle. The movements of the real world are, of course, not uniform like the circles of the Ptolemaic system. As Bronowski puts it, such movements "cannot be analysed until a mathematics is invented in which time is a variable. . . . But the Renaissance had the intellectual equipment: the inner eye of the painter, and the logic of the mathematician."[32]

Johannes Kepler (1571–1630), who took over Tycho Brahe's papers after his death in 1601, was intrigued by what appeared to be wrong with the orbit of Mars around the sun. His discovery that the orbit was elliptical though regular would revolutionize astronomy. But a planet runs along an ellipse at varying speeds. The old mathematics of static patterns and uniform motion would therefore no longer suffice. Kepler did attempt to deal with the problem through a method which he described in his *Volume Measurement of Barrels* (1615). However, as James Burke puts it: "The only problem remaining was the mathematics. Even with Kepler's new geometrical technique, calculation was dauntingly difficult and time-consuming."[33]

The mathematics describing instantaneous motion was invented independently by Isaac Newton (1642–1727) and Gottfried Wilhelm Leibniz (1646–1716). Though their invention is now known (after Leibniz) as the differential calculus, Newton's name of *fluxions* gives a better description of the flux of time that had stopped like a shutter. Bronowski says, of their invention of the differential calculus, that "it is now so familiar to us that we think of time as a natural element in a description of nature; but that was not always so. It was they who brought in the idea of a tangent, the idea of acceleration, the idea of slope, the idea of infinitesimal, the idea of differential."[34]

Newton's clockwork universe is generally considered a relatively static construct by contrast with Einstein's twentieth-century model. Its mathematics is, however, already the mathematics of change and includes the numbers that describe time. For example, Albert Einstein's theory of general relativity uses extensively the calculus of variations. When he began to study the calculus some fifty years ago, this writer little thought that the word derives from the Latin *calculus*, the pebbles used in early forms of mathematics. Nor did he realize that "dv" divided by "dt" (an infinitesimal part of velocity over an infinitesimal part of time) marked the entry of time as a fourth dimension in the new mathematics of change. We have earlier noted that the new disciplines that developed during or after the British horological revolution—such as paleontology, archaeology, the dialectics, and evolution—are virtually all energized by a concern with time. And so it was with the new mathematics, beginning with the calculus of Newton and Leibniz.

SPACE-TIME

Though the origins of the idea of space-time may go back beyond our century, this is essentially a modern idea. Geza Szamosi, himself a musician, has suggested that time is a mental construct, and that "these standards . . . were first introduced, studied and used during the late Middle Ages in the theory and practice of polyphonic music." Szamosi argues that writing down the temporal structure of polyphonic melodies involved comparing and finding symbols for durations of time, and regulating its flow. Galileo, whose father was a pioneer in new musical styles, was, like the other students of his time, educated in the quadrivium. This was an integrated unit of study that brought together arithmetic, geometry, astronomy, and music. In Szamosi's words, "By the sixteenth century, this new perception of time [that derived from musical *tempo*] had become thoroughly common, if not yet clearly articulated." He argues further, of Galileo's understanding of time from his musical studies, that "on the strength of his free-fall experiments, [he] distilled this intuition [of time-related values] into a succinct mathematical concept."[35]

For his free-fall experiments, Galileo is said to have used the pulsilogium or pulsometer, invented by his friend Santorio Santorio for timing the pulse. This consisted of a weighted thread, which acted much like a pendulum, including the fact that the length of the thread could be altered. Whether he employed this or a form of clepsydra, as stated by Szamosi, the point is that Galileo could actually experiment by timing the rates at which different balls rolled down various slopes. He no longer needed to hypothesize about the trajectory of a cannonball. From this, Galileo was able to arrive at what would later lead to the "thirty-two feet per second per second" rule. He eventually published his experiments in *Dialogues Concerning Two New Sciences* (1638).[36]

From our point of view, Galileo's free-fall experiments are an early instance, perhaps the first instance, in which a natural phenomenon related to space is measured and related to time. Galileo was wise enough to choose problems in mechanics—such as those related to falling bodies or projectiles—which were simple enough for him to solve. The later developments would not come until Newton (who was born in the year that Galileo died) had produced the mathematics for measuring variable velocities.

The study of dynamics is yet another, though related, area of mathematics in which disciplines concerned with time are introduced in the seventeenth century. In mechanics, for example, Archimedes (287–212 B.C.) had developed systematically the principles of levers and of the statics of fluids. With the advent of the calculus, however, dynamics, as distinct from statics, becomes a possible discipline. During the last three centuries, it has been used in many areas of pure and applied research. As a result of the development of relativistic and quantum mechanics, early in the twentieth century, many mathematicians and physicists have turned away from what we now call classical mechanics.

Galileo had devised a plan for measuring the speed of light by attempting to quantify the time it took for a distant person to see the light of a lantern once its shutter had been opened. The first person to measure the speed of light, which is an important aspect of the concept of space-time, was Ole Roemer (1644–1710), the Danish astronomer. He reported the first real measurement in 1676. In 1679 he met Isaac Newton, Edmund Halley, and John Flamsteed on a visit to England. Roemer had been studying the eclipses of Jupiter's moons through a telescope. He noticed that the intervals between the disappearance of some of the moons behind Jupiter varied in accordance with the distance between Jupiter and the earth. He based his calculations on the assumption that the delay of the eclipse must result from a direct relationship between the extra (or lesser) distance from the earth, and the velocity of light. From this he calculated that the speed of light must be 225,260 kilometers per second, in modern terms. The current value for the speed of light is 299,727 kilometers per second. The main reason for the inaccuracy of Roemer's measurement is that he did not know precisely what the distance to Jupiter was.

The Michelson interferometer (1881) allowed much more accurate measurements for the speed of light. The subsequent Michelson-Morley experiment of 1887—which demonstrated that the earth's motion had no influence on the velocity of light—formed the starting point for Albert Einstein's theory of relativity. The Michelson-Morley experiment was confirmed, in 1926, by R. J. Kennedy, using an instrument with greatly increased sensitivity. Current measurements for the speed of light, using a laser beam and an atomic clock, have reached a very high order of accuracy. The concept of relativity, in its broadest sense, has been with us since the time of Galileo. This aspect of relativity is implicit in the question of whether the laws of nature and the physical situations described by them will always appear to be the same to different observers, whose own states of motion are not necessarily identical. What have been lacking until our own century have been the mathematical and physical instruments with which to undertake measurements.

With respect to discoveries and inventions, we have just noted that the very telescope and moons of Jupiter involved in Galileo's famous experiment were later used by Roemer in a first attempt to measure the speed of light. This should remind us of the curiously symbiotic relationship between inventions and their appearance at the time when they are most needed. There was never so much paper being written upon as in the period just before printing was invented. Hence, spectacles for the scribes appeared from about 1300. But the spectacles were not turned into telescopes until just before they were needed, at the time of Galileo. Similarly, there were never so many horse-drawn carriages as in the period just before horseless carriages were invented; and many more such examples can be adduced.

In the period before cannon on the battlefield, transoceanic voyages, and the explosion of the crystal spheres in the Ptolemaic universe, there was little need for telescopes. Once they were required, the necessary technology, found in spectacles, was at hand. The invention of the chronometer, as we have seen, took rather longer. But the related horological revolution, which came directly out of the needs of astronomy and navigation, has given us, in terms of time measurement, the impetus for virtually all of our new disciplines. These new disciplines contribute, as we shall see in due course, to the industrial revolution.

ECHOLOCATION

Most inventions copy or extend what already exists. Thus printing looked in the first place exactly like handwriting, and the telescope extends the natural eye. Similarly, new inventions, before becoming modified, often follow natural forms. It is, therefore, no accident that submarines look like fish, and airplanes look like birds. Because inventions are sometimes arrived at in great secrecy and under the pressure of great need (necessity having long been the mother of invention), it is not always possible to know for certain whether they derived directly from nature, or developed quite independently. Such is the case with echolocation.

Two of the most important inventions during World War II—which derive directly from the concept of space-time and would further rationalize the measurement of time and distance—were those related to echolocation. During World War II the Allied forces relied heavily on underwater devices for detecting submarines, devices which the British called asdic, and the Americans called sonar. They also relied on the radio echoes, which the British called RDF (radio direction finder), and the Americans called radar. What the sonar and the radar pioneers apparently did not know, though the world knows it well now, was that bats and many other animals have, for tens of millions of years, been using echolocation to measure distance-time. In fact, nature's feats of detection and navigation should have impressed greatly the wartime engineers with their comparatively primitive examples of human ingenuity.

Richard Dawkins relates how Donald Griffin, who was largely responsible for the discovery of sonar in bats, tells what happened in

1940 when he first reported on the new discovery of the facts of bat echolocation. Among the astonished conference of zoologists, "one distinguished scientist was so indignantly incredulous that 'he siezed Galambos by the shoulders and shook him while complaining that we could not possibly mean such an outrageous suggestion. Radar and sonar were still highly classified developments in military technology, and the notion that bats might do anything even remotely analogous [was] . . . not only implausible but emotionally repugnant.'"[37]

Not only do inventions seem to appear when they are most needed, but they often occur in duplicate form. We have noted this in regard to Newton and Leibniz, as well as Darwin and Wallace. Something similar occurred on both sides in World War II, in research into echolocation and atomic fission. So secret was echolocation for the small ships of the Royal Navy, on which I served during World War II, that we blew up one experimental underwater echo-sounding device when we lost it in English waters. Also, in case our ships were captured, we were ordered to remove our asdic before operating behind the Japanese lines in the Arakan, during the campaigns of 1943 and 1944. Such an action seems ridiculous now when so many facets of navigation have been rationalized through the use of echolocation. Today, most private yachts carry depth-sounding gear, and radar systems are widely employed for both aerial and marine navigation.

It is perhaps even more remarkable that Dawkins should be able to point to so many apparently independent forms of "distance-time" echolocation equipment that have since been discovered in nature itself: "At least two separate groups of birds do it, and it has been carried out to a very high level of sophistication by dolphins and whales. Moreover, it was almost certainly 'discovered' independently by at least two different groups of bats." As Dawkins further points out, though no insects and no fish have so far been found to use sonar, "two quite different groups of fish, one in South America and one in Africa, have developed a somewhat similar navigation system" in waters that are too muddy for vision. The fish exploit electric fields in water, a physical principle "even more alien to our consciousness than that of bats and dolphins."[38]

Our present purpose is to consider some of the recent developments concerning echolocation, in particular, and the relating of distance and time, in general. As R. V. Jones observes, "The millionfold or so improvement in precision of time determination over the last fifty years far exceeds the improvement in precision of length measurement in the same period and so it has been worth substituting time for length as the property to be measured wherever possible." He adds that, as a result, the meter is being redefined as "the length of the path traveled by light in vacuum during a time interval of 1/299, 792, 458 of a second." Since the standard deviation of the last figure is about ± 1.2, the uncertainty in this unit of length is about one part in 3,000 million. As Jones indicates, this apparently recondite method for measuring length has been anticipated ever since the use of radar in World War II. For radar, at that time, a range of a hundred miles against an aircraft was considered a good performance. By 1946, however, the first echoes produced by radar were detected from the moon.[39]

During the next twenty or thirty years, echoes were received back from the sun and the planets. Features were also revealed from Venus, for example, with a detail down to five kilometers from a range of about fifty million kilometers. Only two hundred years earlier, in 1761, Harrison's chronometer, when being tested at sea, had given the longitude of Madeira in one direction when the captain's dead-reckoning had pointed in another. The fact that the chronometer proved to be right then seemed to be almost a miracle. However, we have come to expect much more from our time-reckoning devices now that we are moving into interplanetary and even interstellar navigation.[40]

As James Burke puts it, when the lunar module Eagle touched down at 13 hours, 19 minutes, 39.9 seconds U.S. Eastern Standard Time, its 240,000-mile journey had been projected at 4 days, 6 hours, 45 minutes, and 39.9 seconds. The "only reason for needing to know what time it is during a journey is in order to find out where you are." The target was to be reached no more than one and a half miles off center, and all six of the Apollo landings between 1969 and 1972 achieved the required degree of accuracy.[41] But we have been doing better still. As R. V. Jones notes, "The astonishing degree to which it has been possible to navigate planetary probes to Saturn and Jupiter and, hopefully, now to Uranus, has depended on these radar measurements, and, even though light travels at about 186,000 miles a second, it takes nearly three hours to send out a pulse and get it back from Saturn!"[42]

Probing beyond the moon has required an even more accurate clock. The hydrogen maser, which was to serve this purpose, was first made at Harvard in 1960. As Marcia Bartusiak puts it, "As long as man strayed no farther than the moon, it was simple enough to determine the distance of a satellite or a spacecraft. Scientists aimed a radio signal (which, like light, travels at 186,000 miles per second) at the craft and used a quartz clock to measure the time it took the signal to reach the target and return to earth." For the present much greater distances, they use a technique in which "two or more widely separated radio telescopes, synchronized with maser clocks, look simultaneously at the same object." In order to track Voyager, for example, "Clock A times arrival of the spacecraft signal at first antenna; a tiny fraction of a second later, the signal arrives at Clock B." Since one knows both the time delay and the baseline distance, experts can compute an angle that locates Voyager.[43]

But maser clocks can also be used for making very precise measurements on the surface of the earth. Some tests have already been carried out by astronomers at the Owens Valley Radio Observatory in California, the Haystack Observatory in Massachusetts, and the Jet Propulsion Laboratory in Pasadena. They use radio telescopes trained on a quasar, which, being several billion light years away, can be considered fixed in the sky. The difference in the arrival times for the quasar's radio emissions at the Owens Valley and Haystack observatories are measured by maser clocks. This information—together with the knowledge that the length of the baseline on earth is 154,680,381 inches—provides the data required for deciding whether there has been any expansion or contraction of the earth's crust along the baseline. The results indicate that the North American plate has remained steady over the five years leading up to 1987, give or take an inch. During one eleven-week period, however, the 209-mile distance between Owens Valley and Jet Propulsion, across the San Andreas Fault, grew by about eight inches. It is estimated that one twenty-four-hour period of such testing produces more data than the Internal Revenue Service handles in a year.[44] Through the progressive rationalization of the measures of time and distance, as well as the achievement of very high levels of accuracy, men and women have certainly come a long way since the days of finger counting.

SPACE-TIME, DISTANCE-TIME, AND NUMBERS "MEASURELESS TO MAN"

For dealing with echolocation and associated forms of the measurement of length through time, I have used the term *distance-time*, in order to differentiate it from Hermann Minkowski's related *space-time*. In our own century it has become less possible to separate time from space. Even when two people view the same object in the same room, we know that they do not view it simultaneously. This question was first considered by Minkowski in 1908. As he maintained, "From now on the ideas of space and time as independent concepts shall disappear and only a union of the two shall be retained as an independent concept."[45] Samuel Alexander wrote, as early as 1920, that "a world in Space and Time which was formulated by mathematical methods by the late H. Minkowski . . . had been used or implied in the memoirs of Messrs. Lorentz and Einstein, which along with Minkowski's memoir laid the basis of the so-called theory of relativity. . . . Every point has four co-ordinates, the time co-ordinate being the fourth. Hence it follows, as Minkowski writes, that geometry with its three dimensions is only a chapter in four-dimensional physics."[46]

Thus it may be taken that any event which occurs in the world is determined by the space coordinates x,y,z, and the time coordinate t. But Alexander cautions us that "Time does with its one-dimensional order cover and embrace the three dimensions of Space, and is not additional to them. To use a violent phrase it is spatially not temporally voluminous." And he adds that "Space, even to be Space, must be temporal. . . . It follows also that the three dimensions of Space, just because they correspond to the characters of Time, are not in reality independent of each other."[47]

The geometrical interpretation of the Special Theory of Relativity by Minkowski was of the most far-reaching significance. It provided the clues that enabled Einstein to generalize the theory and thereby include gravitational forces. J. T. Fraser takes the story further: "For Einstein the mental grasp of things meant the Pythagorean geometrization of time into space-time. It is only a small step from here to C. W. Misner, K. P. Thorne, and J. A. Wheeler who go even further when they maintain that 'the proper arena for the Einstein dynamics of geometry is not spacetime, but superspace.'" As Fraser adds, "A cut through superspace is a 'leaf of history,' which 'describes the deterministic, dynamical development of space with time.'" He deals further with superspace in *The Genesis and Evolution of Time*.[48]

For measures of time and space that are relative to the eye of the beholder, Jacob Bronowski's chapter "The Majestic Clockwork" is particularly helpful. He highlights the change in attitudes toward time: "For Newton, time and space formed an absolute framework.... His is a God's eye view of the world: it looks the same to every observer, wherever he is and wherever he travels. By contrast Einstein's is a man's eye view, in which what you see and what I see is relative to each of us, that is to our place and speed." And this relativity cannot be removed, because we can only communicate at the speed of light and not instantaneously. The paradox then is that time is relative. It is different for the traveler, for example, and for the person who stays at home. Consider what would happen if a man could travel on a beam of light, leaving home at noon for a journey of 186,000 miles. In looking down at the clock he had left one second previously, it would still be at noon, because "it takes the beam of light from the clock exactly as long [to make the journey] as it has taken" the traveler. Or as Stephen Hawking puts it, "the discovery that the speed of light appeared the same to every observer, no matter how he was moving, led to the theory of relativity—and in that one had to abandon the idea that there was a unique absolute time.... Thus time became a more personal concept, relative to the observer who measured it."[49]

We now measure distances in the universe by light-years. Since the universe is thought to be about fifteen billion years old, Bishop Ussher's six thousand years seem minute. Dawkins is able to argue against Bishop Ussher's chronology by using the language of light-years: "It is incompatible, not only with orthodox biology and geology, but with the physical theory of radioactivity and with cosmology (heavenly bodies more than 6000 light-years away shouldn't be visible if nothing older than 6000 years exists; the Milky Way shouldn't be detectable, nor should any of the 100,000 million other galaxies whose existence modern cosmology acknowledges)."[50]

Robert Vessot, of the Harvard-Smithsonian maser group, has already rocketed a ninety-pound maser clock 6,200 miles above the earth. The gravity at that height is weaker than on the surface of the earth. The test has therefore been able to provide additional verification for another of Einstein's predictions, namely that time speeds up in a weaker gravity field. In due course, Vessot's group hope to send a forty-pound clock perhaps as far as Jupiter. In this way, they should be able to verify by measurement another of Einstein's predictions related to space-time. That prediction concerns the so-called "ripple in space-time," which should result in an effect on clocks when the gravity wave from a supermassive black hole passes over them. They hope to measure how the wave varies the frequency of a clock on earth, as well as one in deep space.[51]

The theory of relativity would seem to suggest that space-time is relative to the eye of each particular beholder. That would not have been possible in the Ptolemaic model when the earth was at the center of the universe and hell was at the center of the earth. Thus one used to move outward from imperfection at the center to perfection beyond the *primum mobile*. By way of contrast, the Copernican revolution would appear to have changed the universe to one in which no part is more privileged than any other. In Bertolt Brecht's *Galileo* the Very Old Cardinal questions the possibility of the new cosmology: How could God place such a miracle as man, "such a masterpiece, on a little remote and for ever wandering star? Would He have sent His Son to such a place?"[52]

THE ATTEMPT TO BRING MAN BACK
INTO THE PICTURE

Such ideas as relativity, the possibility for intelligent life in other parts of the cosmos, and even the concept of the evolution of inorganic forms and organic matter derive ultimately from the Copernican revolution. Tony Rothman discusses such aspects of the new cosmology in "A 'What You See Is What You Beget' Theory." Rothman notes that "the term Anthropic Principle was coined in 1974 by cosmologist Brandon Carter . . . in accord with the fundamental rule of physics that you must take into account the properties of your experimental apparatus, in this case us." At the end of Rothman's article, a character on Earth says: "The conditions we observe in the universe are those necessary to give rise to intelligent life, otherwise *we* wouldn't be here to observe them." From one star comes the reply: "Oh that's great. He thinks. . . ."

And from another star the line is capped with: "... therefore, we are."⁵³ Needless to say, this is a conflation of Descartes and Berkeley.

What such ideas do demonstrate is that seventeenth- and eighteenth-century philosophers were heading directly toward the so-called Anthropic Principle, even without realizing what it was going to be called. Johann Gottlieb Fichte's *Wissenschafftslehre* (1794) had reversed the epistemology of his teacher, Immanuel Kant. Now, instead of the eye assimilating through the senses information about the external "Ding an sich," the eye became the actual creator of the external phenomenon. Though there were differences, it was almost as though Alhazen had never existed. For his early use of the Anthropic Principle, poor Fichte was accused of the heinous crime of atheism. His influence on the English and German Romantics was, however, considerable. Wordsworth's formulation, in *Tintern Abbey*,

> ... of all the mighty world
> Of eye and ear—both what they half create,
> And what perceive ...

indicates the nature of this influence, just as does Coleridge's "esemplastic" quality of the poet's mind in *Biographia Literaria*.⁵⁴ The fact that Wordsworth thought of himself as a high priest of poetry is not perhaps as presumptuous as one might think at first. The Greek *poetes* does in fact mean "creator," though admittedly spelled with a small *c*.

We may suspect that, however unwittingly, God has always been made in the image of man. But it can only be in the last three hundred years that some may have begun to wonder whether he might not be made in the image of man, the scientist. One of the problems of the new universe with which scientists are now grappling is that their universe requires a model beyond our capacity to envisage. Its very measurements are, and perhaps always will be, beyond our ability to quantify. Science requires, more than anything, the power of prediction that comes through accurate measurement. Indeed, sometimes, as Stephen Hawking puts it, "we have increased the sensitivity of our measurements ... only to discover new phenomena that were not predicted by the existing theory."⁵⁵ Thus far, in this study, we have been dealing with the rationalizations that have led to our being able to take measurements that are more and more precise. But even a model for the new universe is not yet within our grasp.

The model for the universe most generally accepted at present is still patterned on an expansion (though admittedly a gigantic one) of the Judeo-Christian linear historic eschatology of finite time. In this model there was a big bang, say fifteen billion years ago, from which the universe will expand indefinitely. But without an eternal God who can live on beyond the finite lives of individual scientists, how are we to explain either the beginning or the ending of the universe? If there really is an Anthropic (or Anthropomorphic) Principle, does it die when the last scientist dies? Such speculation is already leading in the direction of what is called a "closed universe." In that model, the rate of expansion (known as the Hubble constant) might have been slower than is now thought, and the density of mass and energy that can be found in the universe might be greater than we presently believe. If that is true, the universe will, at one point, cease expanding and start to collapse. Should the universe collapse into a gigantic black hole, the process could start all over again, and again, and again. This would surely give us the more satisfying cyclic model of the eternalistic eschatologies. Ironically, that model would also be much more in accord with the large numbers that have given us the Hindu-Arabic numerals, and which we now use for measuring the universe.

Although the expanding universe was implicit in Einstein's General Theory of Relativity, Hawking tells us that he was still "so sure that the universe had to be static that he modified his theory to make this possible." For that purpose, he introduced a so-called cosmological constant, which he was later to declare the biggest mistake of his life. The nonstatic universe was predicted in 1922 by Alexander Friedmann, the Russian physicist and mathematician. In that decade, as Hawking puts it, "astronomers began to look at the spectra of stars in other galaxies." Since they discovered that "their spectra shifted toward the red end of the spectrum," it was clear that, in accordance with the Doppler effect, the galaxies were all moving away from one another. Moreover, as Edwin Hubble indicated in a paper published in 1929, the farther a galaxy is the greater its "red-shift," and therefore the faster it is moving away. As Hawking tells us, in 1965 Roger Penrose argued that when a star collapsed under its own gravity, "the density of matter and the curvature of space-time become infinite. In other words, one has a singularity contained within a region of space-time known as a black hole." Hawking realized that if he reversed the direction of time in

Penrose's theorem to that applicable in our present expanding universe, it must have begun with a singularity and commenced its expansion through what we now call the "big bang." This resulted in a joint paper by Hawking and Penrose in 1970. Hawking, however, now feels that a new theory should be developed which will also take into account the uncertainty principle of quantum mechanics.[56]

Generally, a good model, or a formula like $E = mc^2$ (where E is energy, m is mass, and c is the speed of light), is also a simple one. But Einstein spent the latter part of his life in the unsuccesssful search for unified field theories. A widespread current search for a unifying theory deals with what are known as "strings." It has been named TOE, or Theory of Everything. The string theory went far beyond the four dimensions of the space-time of Minkowski and Einstein. In its primitive stage, as Gary Taubes tells us, "it was couched in the mathematical terms of a universe with 26 dimensions." More recently, the scientists seem to be in the process of rationalizing their theories. Taubes explains further that "the focus of the work now being done by physicists (with a lot of help from friendly mathematicians) is to bring superstring theory down from its ten-dimensional ivory tower to a four-dimensional form that can be used to make predictions that might be confirmed or refuted by earthly experiments."[57]

Though there may well be more things in heaven and earth than are dreamt of in our philosophy, it would seem likely that an elegant theory should not require ten, let alone twenty-six, dimensions. Be that as it may, some mathematicians have used our newly rationalized Hindu-Arabic numerals to move out into theories beyond the comprehension of mere mortals. Yet this certainly does not mean that our rationalization of numbers into the decimal system has been in vain. The level of communication through numbers is now far higher and far more widespread than it has ever been. As Ifrah puts it, in making a comparison with the fifteenth century, "The greatly increased facility with which the average man today manipulates number [sic] has often been taken as proof of the growth of the human intellect. The truth of the matter is that the difficulties then experienced were inherent in the numeration in use."[58]

There was certainly some reaction to the rationalization related to evolving into the decimal system, as distinct from base-twelve, -twenty, and -sixty numbers. But this was not of the same order as the reaction to the rationalization of the calendar, of chronology, or of weights and measures. Such reaction as there has been has led to the retention, thus far, of part of the duodecimal and sexagesimal systems in numbers related to time and the division of the circle. In general, however, people do not seem to have considered the rationalization of numbers a major threat to their religious ideologies or to their economic well-being.

A generation before the period with which we are mainly concerned, Francis Bacon, the father of English science, wrote a plan for the first research institute. This appeared in the description of "Solomon's House" in *The New Atlantis* (1626). Bacon's plan was consciously used as the model for the Royal Society (1662), which was granted its charter by Charles II within some two years of the Restoration. In Bacon's plan he clearly foresaw the difficulties inherent in the relationship between men of science and men with political power. Bacon says, in a surprisingly relevant passage not often quoted, "And this we do also: we have consultations which of the inventions and experiences which we have discovered shall be published, and which not; and take all an oath of secrecy for the concealing of those which we think fit to keep secret; though some of those we do reveal sometimes to the state, and some not."

Over three hundred years later, the rationalization by scientists that has given us space-time and distance-time has met with an ambivalent reception. There is a pride in the achievements of the physicists and mathematicians, and a greed for the valuable spin-offs that derive from their work. But there is also a very real and widespread fear of the hydrogen bombs and the ballistic missiles that come directly out of this research. In one sense, man, the scientist, certainly has taken the place of God. Until the advent of the applied research that derived from space-time and distance-time, God was the sole controller of Armageddon. Today, man can achieve this without external intervention.

VI The Ascendancy of English

There were certainly no hints from its beginnings that English would become the dominant language of science, business, and politics in our global village. The Angles—from whom we derive the name of our language—first landed in Britain in A.D. 449. They pushed out the Celts, whom they called foreigners or *wealas*, and from this we derive the term *Wales*.[1] Though English has borrowed remarkably few words from the Celtic, much of the Anglo-Saxon heritage remains in our language. Most of our everyday words, including agricultural words, come down from the Old English of the Angles and Saxons.

Christianity entered Britain—first from the south under the leadership of St. Augustine, and then from the north under Aidan from Celtic Ireland—in A.D. 597 and 635 respectively. This resulted in an infusion of Latin, and of the more sophisticated ideas that its words were accustomed to convey. It also gave to English the valuable and peculiar flexibility that it derives from being able to create new and often more abstract meanings by interchanging nouns, adjectives, and verbs almost at will. In English-speaking countries we not only have tables and chairs, but also a chairman who can chair a meeting at which a previously tabled motion is tabled once again.

Alfred the Great (849–99) was the first king who consciously supported Old English as a language. He arranged for a number of key books to be translated from the original Latin, and he also translated some himself, including Boethius's *Consolation of Philosophy*. It was his intention to provide vernacular translations of "some books which may be most necessary for all men to know." After Alfred, the Saxons lived in relative amity alongside the Danes, who had been invading, mainly from the north, since about A.D. 750. Because the languages of both peoples were Germanic in origin, they could understand each other sufficiently for a pidginization to take place. As a result, Old English became very much simplified, instead of remaining a heavily inflected language.

The Norman invasion of 1066 affected English much more radically than had the invasions of the Danes. The Normans, or Northmen, had themselves, like the Danes, been Vikings. Three generations before the conquest of England, they had taken over Normandy (North-man-dy). Now their descendants, of mixed Scandinavian and Frankish ancestry, brought their Norman French into England. At first Old English remained a separate language of the subservient indigenous people. But this did not last long. Through a process of intermarriage, analogous with what had occurred earlier in Normandy, a new language evolved. For the period 1150–1500, we call this language Middle English.

It is noteworthy that in the case of both the Normans who invaded Normandy and their descendants who later invaded England, it was the language of the indigenous people that prevailed. It did so, however, in a somewhat altered form. In the case of Middle English, this involved yet another valuable infusion of words as the two peoples slowly became integrated. As early as 1100, Henry I addressed "all his faithful people, both French and English, in Herefordshire."[2] The process of intermixing reached a turning point in 1244, when the king of France insisted that, since no man could serve two masters, all men must decide whether their allegiance was to the king of England or to himself.

Yet, for many years in England, three languages operated alongside one another. Though, in theory, French was spoken by the knights at court and in government, soon enough the descendants of the Normans would be having to learn their French rather than acquiring it from the cradle. In theory, too, Latin was the language of the church, but one could only communicate with the common people in Middle English. Thus it was that Chaucer (ca. 1340–1400 A.D.), the son and grandson of wine merchants, found himself faced with the choice of whether to write his poetry, for reading to the court, in French, English, or Latin. He chose well. In 1381 Richard II spoke in English to the peasants during Wat Tyler's rebellion. When Richard II was murdered, in 1400, Henry IV both claimed and later accepted the throne in English. Chaucer deals sharply with those who aspire to speaking French when he undercuts the Prioress in the General Prologue of *The Canterbury Tales*:

> And Frenssh she spak ful faire and fetisly,
> After the scole of Stratford atte Bowe,
> For Frenssh of Parys was to hire unknowe.[3]

ENGLISH: RATIONALIZING A SCAVENGER TONGUE

Chaucer was also very well aware of the differences of words, accents, and spellings within the English tongue. He alludes to this at the conclusion of *Troilus and Criseyde*:

> And for ther is so gret diversite
> In Englissh and in writyng of oure tonge,
> So prey I God that non myswrite the,
> Ne the mysmetre for defaute of tonge.[4]

The last two verses are particularly prescient. Two months before his death, in 1700, John Dryden published *Fables Ancient and Modern*. In this he translated parts of Chaucer, whom he calls "the Father of English poetry." As Dryden puts it, "The verse of Chaucer, I confess, is not harmonious to us. . . . for he would make us believe the fault is in our ears, and that there were really ten syllables in a verse where we find but nine."

The Norman tongue may have died out in England, but it enriched Middle English by no less than ten to twelve thousand new words.[5] A similar number of words was again added during the English Renaissance. By this time, however, the reasons were manifold. The new imports included words derived from Latin and Greek by the early scientists and words brought in by Elizabethan seamen and Protestant immigrants in the period after Elizabeth's succession. It is estimated that between 1500 and 1640, some twenty thousand items—such as pamphlets, folios, broadsheets, and Bibles—were printed in English.[6] This can be accounted for both by an increase in literacy and by an increased pride in the mother tongue. Sir Philip Sidney concludes his argument in *An Apology for Poetry* (1595): "Lastly, our tongue is most fit to honor poesy, and to be honored by poesy." It was already the language of Shakespeare, who had some thirty thousand words at his command.

Before they were rationalized through the regular use of dictionaries, words were spelled phonetically in an unbelievable number of ways. In 1845 Joseph Hunter gave no less than twenty-six spellings for Shakespeare's own name, "all taken from writings of nearly the poet's own age."[7] Marlowe's name was spelled in seven different ways, and so was Raleigh's. Much the same was true of the later Edmund Halley (1656–1742), of comet fame. His name was spelled Halley, Hailey, Haily, Hally, and Hawley, during his own lifetime. If we take at random some of the terms with which we have been dealing in this study, the *Oxford English Dictionary* shows that, over the centuries, *dozen* has been spelled in thirty-one, *fathom* in twenty-nine, *ounce* in sixteen, and *hour* in eighteen different ways.

The chaotic nature of English spelling is still an embarrassment for which much is owed to the early printers. The first English printer—William Caxton (ca. 1421–91), a mercer and diplomat who learned printing at Cologne—was well aware of the problems related to spelling. Late in his life, he set up his press in Westminster, and became a translator, writer, editor, and publisher. After publishing the first book in 1477, he went on to print many works, including two editions of Chaucer. In Caxton's time spelling tended to be phonetic, and there were considerable regional variations in words and accents. The fact that he used the London accent—he had been apprenticed to Robert Large, who became the lord mayor of London—may well have affected the nature of Standard English to this day. Caxton's story about the difficulties with language in his time—as well as the idiosyncratic nature of his own syntax and spelling—can help us to understand the formidable predicaments of the printer's trade:

> And one of theym named Sheffelde, a mercer, cam in-to an hows an axed for mete and specially he axed after eggys. And the goode wyf answerde that she coude speke no Frenshe. And the marchaunt was angry, for he also coude speke no Frenshe, but wolde have hadde egges; and she understode hym not. And thenne at laste another sayd that he wolde have eyren; then the good wyf sayd that she understod hym wel. Loo! what sholde a man in thyse dayes

now wryte, 'egges' or 'eyren'? Certaynly it is harde to playse every man by cause of dyversite & change of langage.... But in my judgemente the comyn terms that be dayli used ben lyghter to be understonde than the olde and auncyent Englysshe.[8]

The diversity of the English vocabulary is probably its greatest strength, owing very much to the fact that it is a scavenger tongue. But, having borrowed words from virtually every language in the world, it is plagued with inconsistencies. In addition, during the late seventeenth century, a whole host of technical terms were consciously taken out of the workshops and included in the written language. As a result, English has a vocabulary of some 500,000 words, along with a similar number of technical and scientific terms that remain uncataloged. By way of contrast, the Germans have about 185,000 words, and the French—who since the Académie Française was founded in 1634 have regularly attempted to eliminate foreign borrowings—have fewer than 100,000 words.[9] In the area of the metric system, their logic and sense of rationalization have served them well, but these very qualities seem to have lost them the world leadership in language that they have so earnestly desired.

Curiously, the melting-pot nature of English society in the seventeenth and eighteenth centuries has been reflected in a laissez-faire attitude toward the vocabulary. The same can be said of the Americans, who have since become responsible for the further evolution of the English language. The bastard and scavenging nature of the English and their language is colorfully described by Defoe with both pride and exasperation in his *The True-Born Englishman* (1701):

> ... ev'ry Nation that her Powers reduc'd
> Their Languages and Manners introduc'd.
>
> . . . .
>
> *Norwegian* Pirates, Buccaneering *Danes*
> Whose Red-hair'd Off-spring ev'ry where remains:
> Who join'd with *Norman-French* compound the Breed
> From whence your *True-Born Englishmen* proceed
>
> . . . .
>
> We have been *Europe's* Sink, *the Jakes* where she,
> Voids all her Offal Out-cast Progeny.

> . . . .
> *Dutch, Walloons, Flemmings, Irishmen* and *Scots,*
> *Vaudois,* and *Valtolins,* and *Hugonots,*
> In Good Queen *Bess's* Charitable Reign,
> Supply'd us with three hundred thousand Men.
> . . . .
> Thus from a Mixture of all Kinds began
> That Het'rogeneous *Thing, An Englishman.*[10]

Like so much else with which we have thus far been concerned in this study, the language of "That Het'rogeneous *Thing, An Englishman*" became ordered and rationalized during the Restoration and eighteenth century. It is one of those interesting coincidences that the Augustan or neoclassical period in English literature is almost concurrent with the British horological revolution of 1660–1760. The Augustan period reflects a conscious turning toward order after the internecine wars of the Interregnum. The horological revolution satisfied national pride, and it also provided the best possible symbol for order both in the universe and in society. Among philosophers, for example, virtually every major writer and a host of minor ones used the clock analogy to illustrate their central ideas.[11]

Just as with Richelieu's Académie Française in seventeenth-century France, there was a constant yearning for order in the English society of the Restoration and eighteenth century. One even sees this in the way that both nations subscribed to the idea of control through the unities of time, action, and place in the drama. But the British differed from the French. Even their demand for the three unities was not absolute. More important, they damned the use of only a few words, and even then not essentially because they were foreign. What the British did do was to regularize their punctuation, spelling, capitalization, and syntax. They certainly made no effort to eliminate the many duplicate words, which have given an unrivaled richness to their language. They also made no attempt to revert to the various genders—masculine, feminine, and even neuter—which have remained the bane of so many foreign tongues. The appearance of dictionaries, and particularly Johnson's *Dictionary* (1755), regularized the language. Over a period, the many variant spellings were eliminated, and Johnson's *Dictionary* also helped to "fix" the meanings of words by quoting them in context.

THE HOROLOGICAL REVOLUTION
AND THE SIMPLIFIED ENGLISH PROSE STYLE

Critics have long recognized that in the third quarter of the seventeenth century, English prose style suddenly became modern. It is with this dramatic change in the language, during the period that coincides with the British horological revolution, that we must now concern ourselves. As Marjorie Nicolson argues, whatever Swift may have thought of Cartesian and other philosophies, his own language, like that of his age, had been radically affected by Descartes in particular and the scientific Moderns in general. The decisive change toward modernity in English prose style during the last half of the seventeenth century must be credited to them. Not merely Addison, Swift, and Defoe, but even Dryden, Boyle, and Bishop Sprat, writing in the third quarter of the seventeenth century, no longer seem to live in the same world as had Robert Burton and Sir Thomas Browne of the previous generation.[12]

R. F. Jones, in a series of articles written in the 1930s, demonstrated the impact of the scientific movement on the English language. In "Science and the English Prose Style in the Third Quarter of the Seventeenth Century," he dealt with the movement to eliminate unnecessary metaphors and provided the excellent example of Joseph Glanvill's *Vanity of Dogmatizing* (1661), republished in different forms, first as *Scepsis Scientifica* (1664), and then as the first of seven articles in *Essays on Several Important Subjects* (1676).[13] Jones points to the progressive development of Glanvill's style toward the "mathematical plainness" in language being demanded by Bishop Thomas Sprat in his *History of the Royal Society* (1667). The *Scepsis Scientifica* was written specifically to please the Royal Society.

Despite the general caveat on florid language, Sprat made some concessions "to our wits and Writers."[14] One of the essential similitudes of which scientists and philosophers constantly made use was the clock analogy. We find it in Sprat's *History of the Royal Society*, and, significantly, the clock analogies are among the few that Glanvill retains.

The need of horology and astronomy for a "plain" language should be seen as part of a general movement inspired by science. In England, Bacon and his followers were well aware of the advantages of a simplified language. Not only "hard" words, but also unnecessary metaphors belonged to the so-called "florid style" that men like John Webster and Robert Boyle wished to eliminate. Sprat's *History of the Royal Society* contains several such references. He feels that "*eloquence* ought to be banish'd . . . as a thing fatal to Peace and Good Manners." His Royal Society has "indeavor'd to separate the knowledge of *Nature*, from the colours of *Rhetorick*, the devices of *Fancy*, or the delightful deceit of Fables." Indeed, "these specious *Tropes* and *Figures* . . . are in open defiance against *Reason*," and being in league with "*the Passions*: they give the mind a notion too changeable, and bewitching, to consist with *right practice*."[15]

The relationship between figures of speech and lack of order should be noted. What Sprat calls for are writers of an ordered language, "bringing all things as near the *Mathematical plainness*, as they can: and preferring the language of Artizans, Countreymen, and Merchants, before that, of Wits, or Scholars" (italics added). One can see why the aristocratically oriented Augustans had some difficulty in openly accepting such a doctrine. However, Sprat feels that a few obscure and meaningless terms—"such as *Matter*, and *Form*, *Privation*, *Entelechia* and the like"—may be lost, but practical "*Inventions, Motions*, and *Operations* will succeed in place of words." He looks for the return of a time "when men deliver'd so many *things*, almost in an equal number of *words*."[16] This desire for the concrete in language led Swift, in *Gulliver's Travels*, to mock the philosophers of Laputa who, in order to express themselves "by *Things*," carried a great "Bundle of *Things*." He adds slyly, "Another great Advantage proposed by this Invention, was, that it would serve as an universal Language."[17] Swift's mockery of scientists demonstrates that while he owed them a great deal for the simplicity of his own language, he abhorred some of their philosophy, particularly when it was taken to extremes.

In dealing with the influence of the British horological revolution on Augustan literature, in chapter 8 of *Clocks and the Cosmos*, I argue that its impact is complicated by the fact that it influenced both the literary Ancients and the scientific Moderns. In its role as the leading element of the technology of the age as well as the handmaiden of astronomy, horology was connected with the important movement to regularize and simplify the language in the interests of science.[18] The ultimate result of this movement to rationalize is that some 80 percent of all scientific papers are today written in English, rather than in Latin, French, or Russian. But the scientific questioning of which the

༺༒༒༒༒༒༒༒༒༒༒༒༒༺

TO

Alexander Pope, *Efq;*

OF

Twickenham in the County of *Middlefex.*

AS You have been long an intimate Friend of the Author of the following *Poem,* I thought you would not be difpleafed with being informed of fomeparticulars, how *he* came to write it, and how *I,* very innocently, procured a *Copy.*

It feems the D——n, in converfation with fome *Friends,* faid, he could guefs the difcourfe of the World concerning his *Character* after his Death, and thought it might be no improper *Subject* for a *Poem.* This happend above

a

DEDICATION.
a Year before he finifhed it; for it was wtitten by fmall pieces, juft as *Leifure* or *Humour* allowed him.

He fhewed fome Parts of it to *feveral Friends,* and when it was compleated, he feldom refufed the fight of it to any *Vifiter:* So that, probably, it has been perufed by *fifty Perfons*; which, being againft his *ufual Practice,* many People judged, likely enough, that he had a defire to make the People of *Dublin* impatient to fee it *publifhed,* and at the fame time refolved to *difappoint* them; For, he never would be prevailed on to grant a *Copy,* and yet feveral Lines were *retained* by *Memory,* and are often repeated in *Dublin.*

It is thought, that one of his

Servants

Figure 6. Swift's typical dedication to an early version of "Verses on the Death of Dr. Swift" clearly lacks today's uniformity of capitalization, punctuation, typeface, spelling, and syntax. (*The life and Genuine Character of the Reverend Dr. S[wif]t* [London, actually Dublin: J. Roberts, 1733]; courtesy of McPherson Library, University of Victoria)

horological developments were a significant part also coincided with the weakening of religious and feudal bonds. In this respect horology was able to repay its debt to astronomy by providing such essential new models for denoting order as the Cartesian clockwork animal and the Newtonian clockwork universe. Language, too, not only was simplified during the horological revolution but acquired its own "mechanical" qualities. Though he attempted to deny it, Johnson's *Dictionary* helped to "fix" the English language. In addition, this marked the conclusion of a century during which spelling, syntax, and punctuation became relatively ordered and therefore "modern" (figure 6).

The scientific need for simplicity and clarity in language conformed happily to neoclassical values. Dryden's *Art of Poetry* brought into England the Augustan standards of Horace's *Ars poetica,* via Boileau's *Art poétique.* After a passage that uses the much-worked term *reason* in four lines out of the eight, Dryden berates writers who are irregular in their verse and florid in their language:

> Most writers mounted on a rusty muse
> Extravagant and senseless objects chuse;
> They think they err if in their verse they fall
> On any thought that's plain and natural.[19]

Pope's views in the *Essay on Criticism* are well enough known, but a prose passage is worth quoting. In it Pope speaks at least for the ideals of the age that both Ancients and Moderns could share: "Flowers of rhetoric in sermons and serious discourses are like the blue and red flowers in corn, pleasing to those who come only for amusement, but prejudicial to him who would reap the profit from it."[20]

The best Augustan poetry is certainly not rigid or mechanical, despite the fact that the Romantics compared it with clockwork. But for a time the need for order in society—as well as for order and clarity in philosophy and science—made an impact on Augustan poetry that is unique in the history of English literature. Though Dryden, Pope, and others did argue that genius must be allowed to "*snatch a Grace* beyond the Reach of Art,"[21] they did so at a time when there was a real risk of the creative artist's being stifled by the rules. Reason, plainness, and order, if taken to the extreme, could be just as devastating for poetry as the irrational, the obscure, and the chaotic.

Though Defoe ran no risk of being stifled by Augustan rules, he was equally concerned with the clarity and simplicity of the English language. In his *Essay upon Projects* (1697) he argues at length for the need "to polish and refine the *English* Tongue, and advance the so much neglected Faculty of Correct Language, to establish Purity and Propriety of Stile, and purge it from all the Irregular Additions that Ignorance and Affectation have introduc'd." His suggestion, that this be undertaken in a Royal Academy under the patronage of King William, he considers "the most Noble and most Useful Proposal in this Book."[22] Thirty years later, in *The Complete English Tradesman*, Defoe not only urges tradesmen "to write a plain and easy stile," but adds that "easy, plain, and familiar language is . . . the excellence of all writing on whatever subject, or to whatever persons. . . . If a man was to ask me what I would suppose to be a perfect stile or language, I would answer, that in which a man speaking to five hundred people, of all common and various capacities . . . should be understood by them all."[23] What Defoe says here has considerable implications for the rationalization of human beings themselves, with which we will be concerned in the final chapter. Though Swift as an Ancient and Defoe as a Modern were hardly socially or even philosophically compatible, they were at one in providing the "easy, plain, and familiar language" that has given us *Gulliver's Travels* and *Robinson Crusoe*.

The modern world generally accepts English as the best medium for concrete scientific description, but this has not always been the case. The present-day student rarely considers poetry more readily understandable than prose. Yet it is a commonplace that Chaucer's poetry had been written in a much more fluent language than the prose in which he attempted to explain the workings of the astrolabe to the ten-year-old Lewis. We can come much closer than this to the beginnings of the horological and scientific revolution in England, and still find that the language is ill-suited to technological description. The considerable use of Latinate words did little to help.

Walter Charleton is probably as good an example as any of the generation that grew up in England under the twin influences of Baconian science and Cartesian mechanism. He wrote a famous treatise on Stonehenge, received a eulogy from Dryden, and became the president of the Royal College of Physicians. Here, as late as 1654, he describes the history of time measurement:

> *Time is infinitely elder than Motion*, and consequently *independent* upon it. . . . And because the observation of the Suns motion was easie and familiar; therefore did the Ancients invent several instruments, as *Water* and *Sand Hour-glasses*, and *Sun-dials*, and the Neotoricks *Trochiliack Horodixes*, circumgyrated by internal springs, or external weights appensed; and so artificially adequated them to the motion of the Sun, that defines the day by its praesence.

Clearly this language would never do for lucidly communicating the findings of the experimental philosophy. Though he had to describe the fusee as a "pyramid," the crutch as a "little tiller," and the pallets as "little ears,"[24] Huygens still felt it wise to write both his books on the *Horologium* in Latin.

By the end of the horological revolution, technical prose was being published in considerable quantity and was generally lucid for those who understood the technical terms. Just a century after the example from Charleton, a letter about clocks in the *Gentleman's Magazine* says in part: "Methods have already been proposed to correct the errors of the rod of the pendulum, arising from heat and cold; particularly by *Mr. Graham*, in the *Philosophical Transactions*; and also new schemes by *Mr. John Ellicot* F.R.S. published in the same useful undertaking (See Vol. xxiii. p. 429). The last mentioned method I likewise thought of in the year 1748." The influence on literary scholarship speaks for itself. A larger sample would, in addition, demonstrate that punctuation, capitalization, spelling, and syntax were becoming methodized at the same time as parallel developments in technological rationalization. The result is the methodical "matter-of-fact" lucid prose of the modern world. It came out of the laboratories and the workshops at the time when horology—in its dual role of technology's leader and science's handmaiden—had an essential place in the vanguard of change.

Let us look again at the style of Charleton—here writing on the question of time and distance:

> As by the Longitude, of any standing measure . . . we commensurate the longitude of Place: so by the flux of an Horologe do we commensurate the flux of Time. And, insomuch as no motion is more General, Constant and Observed, then [*sic*] that of the Sun: therefore do we assume its motion for a *General Horodix*, by it regulate all our computations, and confide in it as an universal Directory, in our Mensuration of the flux of time.[25]

By way of comparison, here is John Bonnycastle dealing with the longitude in the following century. His language reflects the fact that his is one of the many works of the eighteenth century in which "the most useful and interesting Parts of . . . Science are clearly and familiarly explained":

> At this moment I look at my watch, and find that, instead of its being twelve o'clock by that, it is only nine hours, fifty-eight minutes, forty-four seconds. From this I conclude that, when it is noon at Petersburg, it is before noon at London, and that the difference is two hours, one minute, and sixteen seconds; which, by allowing fifteen degrees to an hour, answers to thirty degrees nineteen minutes.

Since, therefore, the longitude of every place is reckoned from London, and the noon at Petersburg arrives sooner than the noon at London, I know the longitude of that city to be thirty degrees nineteen minutes east of London.[26]

William Derham's *Artificial Clock-Maker* (the earliest editions were 1696 and 1700) was a landmark in technological writing. John Smith's *Horological Dialogues* (1675) had been a nontechnical discussion, and Derham knows that he is not the first to bring the "Art under Rules." But he is conscious of the fact that what others wrote "is understood by very few Workmen. And therefore I have endeavoured . . . to make the matter as plain as I could." Like Robert Boyle, who considered himself an innovator, Derham has learned the trade from men actually involved in it. Not only did he learn about horology from Thomas Tompion and Robert Hooke, but he owes "much to the assistance of L[angley] B[radley], a judicious Workman in *White-chappel*, who drew me up a scheme of the Clock-Maker's Language" (Preface). The new terms and their explanations, which provide the first chapter of the book, offer an excellent example of the way in which technical terms entered and enriched the English language at this time. Toward the end of the *Artificial Clock-Maker*, Derham includes "the historical part" that "hath not been so much as attempted before." The change of subject (which apparently necessitates Latin and Greek quotations) seems to make the same writer regress toward a far more florid and learned style.

Derham's book is generally recognized as the first English horological text of practical value. It also contains one of the first published plates in an English book on horology, though, unfortunately, this includes some fairly obvious errors. Once it had been freed from the inhibitions of the Ancients, technological writing quickly developed the ancillary art of technical illustration. The high standard is perhaps best demonstrated by the *planches* of Denis Diderot's *Encyclopédie* in the second half of the eighteenth century. By then the French were trying to recapture the ground lost to the British during the horological revolution. It is surely significant that more than sixty of Diderot's six hundred folio plates intended to cover Western technology were devoted to horological illustrations, and to the recently improved tools of that essential trade.

THE ATTEMPT TO ELIMINATE JARGON

During the Restoration and eighteenth century, writers consciously set themselves the task of writing in "a plain and easy stile." This involved the elimination of jargon. Naturally, that included the jargon of the few existing and traditional professions, such as divinity, law, and medicine. But the burden on the written English language would increase exponentially, since this was the period that has given us so many new literary forms. Novels, mock epics, biographies, travel literature, periodicals, newspapers, musicals, technological illustrations, dictionaries, and encyclopedias either began during the eighteenth century or were changed in it out of all recognition.

But the horological revolution—which as part of the scientific revolution required a "mathematical plainness" in language—also gave us our modern fixation with time. And that fixation with time has spawned a whole host of disciplines. Herein lies a remarkable irony. A prerequisite for the scientific revolution was a reduction in the hegemony of the guilds and the publication of the processes and terminologies involved in their trades. This movement forced them, as it were, into the public domain. Until the seventeenth century, though never thereafter, trades were called *mysteries*. But today, many of the new disciplines are evolving into new "mysteries," which hide behind a curtain of jargon and "specialization." It is almost as though the jargon of the seventeenth century has returned in a new form and on a vastly greater scale. The title of an article by Edward Tenner attacking "Tech Speak" makes the nature of the new targets clear: "Cognitive Input Device in the Form of a Randomly Accessible Instantaneous-Read-Out Batch-Processed Pigment-Saturated Laminous-Cellulose Hard-Copy Output Matrix."[27]

Often when people have little new to offer, they employ jargon to conceal the paucity of thought. The universities are by no means innocent in this matter. A professor at Brown University is quoted as having the following to say on the subject of familial love: "An altruistic utility function promotes intertemporal efficiency. However, altruism creates an externality that implies that satisfying the conditions as to efficiency does not insure intertemporal optimality." University education faculties throughout North America are particularly adept in the use of such jargon. The standards of education for our children suffer accordingly. But perhaps we should award our tributes more evenly. The University of Hawaii is reported to have offered a seminar on "Isozymes of Glutamine Synthetase Involved in the Processing of Nitrogen Fixed by Rhizobium in Root-nodules of Legumes." Because Swift told us that satire is a glass in which one prefers to see anybody but oneself, I should point also to the jargon inherent in such recent tribes of literary critics as the deconstructionists, semiologists, hermeneutists, and semantic theorists.[28]

Euphemisms involve another important area of jargon that contributes to the concealment of unpleasant facts. Some hospitals, for example, are now referring to death as "negative patient-care outcome." In the area of acronyms, American jargon has grown faster and become more all-embracing than the language itself. In 1672—at the beginning of the modern period with which we have been much concerned—the "Committee for Foreign Affairs" under Charles II was composed of Clifford, Arlington, Buckingham, Ashley, and Lauderdale. From this group—which was the precursor of today's cabinets—was derived the acronym *cabal*. It has since passed into our language. (A similar word had earlier derived from the Hebraic *cabbala*.)

Americans, perhaps influenced by the military, have recently shown a remarkable appetite for acronyms. In 1960 a dictionary of twelve thousand acronyms was published entitled *Acronyms, Initialisms and Abbreviations Dictionary*. The 1985 edition contained some three hundred thousand entries, and the eleventh edition, issued in 1987, has the subtitle "A Guide to More Than 400,000 Acronyms, Initialisms, Abbreviations, Contractions, Alphabetic Symbols, and Similar Appellations." It would seem that a verbal cancer of monstrous proportions has grown out of the misguided attempt to rationalize peripheral and ephemeral organizations and concepts. Those acronyms that have any lasting value—like NASA, UNICEF, and NATO—will undoubtedly enter the language, as cabal has already done. With regard to the vast majority, however, one can only hope that they will follow the cyclic model of the universe and, having eventually sunk back under their own collective weight, will do us all the favor of disappearing.

But change is always frightening and perhaps—since there are few more frontiers through which to develop new words—this is one of the ways in which we will coin fresh ones out of our own language. Certainly, a language that no longer evolves can be expected to die. As early as 1599, Samuel Daniel—who lived in a London newly aware of great seafaring projects—wrote prophetically:

> And who, in time, knows whither we may vent
> The treasure of our tongue, to what strange shores
> This gaine of our best glory shall be sent,
> T'inrich unknowing Nations with our stores?
> What worlds in th'yet unformed Occident
> May come refin'd with th'accents that are ours?
> Or, who can tell for what great worke in hand
> The greatnesse of our stile is now ordain'd?[29]

Twelve years later, the Authorized Version of the Bible was published. It became, together with the plays of Shakespeare, one of the twin pillars by which English literature and language have been supported and, to a great extent, standardized by example.

INTERNATIONALIZING THE ENGLISH LANGUAGE

Samuel Daniel's prophecy has certainly come to fruition. The United States, in part because its early settlers and leaders were English, has been able to impose that language on its own melting pot. When the first American census was taken in 1790, 90 percent of the population were descendants of British colonists. Even though many of the immigrants were not English-speaking, others—like the Irish, the Scots, and the Welsh—had virtually lost their original tongues before they left their homelands. Of the five million people in Ireland in 1800, for example, approximately two million spoke Irish, one and a half million spoke English, and the remaining one and a half million were bilingual. By 1901 English was the only language of 85 percent of the population. By that time, no more than twenty-one thousand in the most remote areas spoke only Irish. Despite great efforts to reverse the trend, the original language is now—as has also occurred elsewhere—at best only a second language for some of the people.[30]

What was true of the United States was also true of English-speaking colonies like Canada, Australia, and New Zealand. Without the ability to speak and eventually to write in English, it became very difficult for new colonists to make a living. Before the advent of radio and television, dialects certainly varied considerably. They usually reflected the area of Britain from which the original emigration had taken place. But two factors above all tended to rationalize accents. The first was that immigrants to the new lands, having traveled so far to get there, generally remained much more mobile within their new countries than their brothers and sisters who were left behind. The second factor that has tended toward standardizing British accents occurred within Britain itself. From the Education Act of 1870 sprang the schools by which people like my wife and me have been educated. They provided, in a rather harsh way, their own melting pot. During the past century this has affected both the society and the speech of middle-class as well as upper-class Britons. The use of the wrong accent provoked the cane of the teacher and the laughter of one's contemporaries—both remarkably strong incentives toward standardization.

From the ages of eight to eleven in preparatory schools and eleven to sixteen or eighteen in the English public school, one was inculcated with the received pronunciation. A by-product of the system was a surprisingly good education, together with other forms of social rationalization approved by the nation and the established church. Thomas Hardy, in *Tess of the d'Urbervilles* (1891), provides an early reflection of the way in which children of the "lesser breeds" might hope to improve themselves. His work mirrors the modifications that occurred as a result of the founding of the National Schools in 1811, and of the passing of the Education Act of 1870. Hardy's description of Tess and her mother, Mrs. Durbeyfield, shows the change that was taking place: "Mrs. Durbeyfield habitually spoke the dialect; her daughter, who had passed the Sixth Standard in the National School under a London-trained mistress, spoke two languages; the dialect at home, more or less; ordinary English abroad and to persons of quality."[31] In Hardy's Wessex, the London mistress was appropriate, because the Received Pronunciation was based on what was generally spoken in the triangle made up by London, Oxford, and Cambridge. It was from this area that the "Queen's English" was developed for exporting—via clerks, administrators, educators, missionaries, and military personnel—throughout that empire on which the sun never set.

Even as late as World War II, it was far easier to receive a commission in the British forces if one's accent and schooling were appropriate. It goes without saying that announcers on the BBC always spoke with such an accent, and for the most part still do. The "lesser breeds," of course, disclosed their backgrounds precisely by not having a rationalized accent. It is possible to argue the case for continuing varieties of English among the one billion people or more for whom that tongue is now either the first or second language. Robert Burchfield has even suggested that English might break down into many different languages, just as the vulgar Latin did after the fall of the Roman Empire.[32] However, the vibrant bastion of English that is now represented by the United States does make such a conclusion unlikely. But the conclusion is even less likely to be reached by anyone aware of the broad spectrum of rationalization with which we have been concerned in this book.

Unless we are prepared to turn our backs on Western technology, the rationalization of many aspects of our lives, including the widespread use of the English language, would seem to be one of the prices that we must continue to pay. In any case, we cannot close our eyes to the standardization that has already occurred. In 1946, just before demobilization from the Royal Navy, I was given my first land-based job. This involved the responsibility for drafting the forty thousand men whose base was the Patrol Service depot in Lowestoft. For the first six weeks, I secretly played the amoral game of speaking to each man who came before me for a minute or two before looking at the card on which his vital statistics were recorded. Despite the fact that I am not particularly observant, I was soon able to train myself, by the process of feedback, to assess a great deal about each individual. I could estimate height within half an inch, weight within two pounds, age within twelve months, and place of birth within fifty miles, or city of birth if it contained at least one hundred thousand residents. After six weeks I stopped playing this game, having become frightened by learning something about the extent to which we are all standardized. Over forty years later, only one criterion would have changed. Even in Britain, physical and social mobility have now resulted in a further standardization of accents. Royalty itself, apart from the queen, is moving in the direction of what is called "down market," and British politicians of the Labour Party have generally upgraded their pronunciation. They only speak with the erstwhile accent of the "working class" when they make an effort to do so. Even English accents are now beginning to lose their distinctions of class.

The greatest inducement for standardizing accents—and this is occurring as much in North America as in Britain—derives from the radio and the television. Just as television commercials tend to level the standard of entertainment to that which may be appreciated by the largest possible audience, the accents too must be relatively standardized. They should not only be understood across the nation, but also be suitable for distribution throughout the English-speaking world. Even Arabs, Russians, South Americans, and Chinese, when they really want to convey a message as widely as possible, tend to be able to do so in relatively standard English.

THE REACTION TO INTERNATIONALIZING ENGLISH

What I am arguing cuts right across the reaction to standardization reflected in the current Romantic "ethnic" revival. That reaction is both valid and understandable, but it does not as yet recognize the price that must inevitably be paid. In Canada the myth of the two founding nations means that all federal forms and all packaging must be in both English and French. The result is that approximately 24 percent of the nation is specially provided with a translation that is some 20 percent longer than it is in English. Yet even in Quebec the people, like their cabinet ministers, know that they must speak English if they want to engage in politics or business on an international level. In Quebec a survey commissioned by the Conseil de la Langue Française discovered to its horror, in 1984, that 72 percent of articles and reports coming out of Quebec university research centers over the previous fifteen to twenty years were in English only. In addition, 54 percent of papers presented at conferences by francophone Quebec researchers and scientists were in English.[33]

In France itself the situation is in some ways even worse. Since December 31, 1975, foreign words (read Americanisms) have been officially declared unacceptable by the Académie Française. The official program of elimination has been supported by successive French presidents. These include Pompidou, Giscard d'Estaing, and even the Socialist, Mitterand. There are "terminology commissions" made up of linguists in nearly all government ministries, a telephone service to provide acceptable alternatives, and a *Dictionary of Official Neologisms*. Naturally, the new words cannot possibly compete with the crispness of American neologisms. *Fast food*, for example, has become *prêt-à-man-*

ger, and *hot money* is now *capitaux fébriles*. With a backlog of eight or ten thousand words in science and technology alone, the work of the Académie is turning into a nightmare. Perhaps they should remember that language, like water, has a tendency to insist on finding its own way. The French might have done better to rely on their own term laissez-faire, and to have learned how to welcome the potential enrichment of their language.

The United States is currently confronted with a problem comparable to that which Canada has faced with its French minority in Quebec. The black minority, like other members of the melting pot, is at last being assimilated into the majority and has made considerable contributions to its adopted language.[34] However, a number of the more recent Hispanic immigrants are attempting to resist the long tradition of linguistic assimilation, which goes back far beyond the involuntary and inhuman conditions suffered by some nine million blacks on the "middle passage." Thus far, in the United States, English has been adopted by the largest and most diversified melting pot in human history. It is estimated that 4.1 million Irish came to the United States in the period 1830–50, 5 million Germans in the 1840s and 1860s, 2 million Italians in the 1860s, 1.7 million Scandinavians in the 1870s, and 3 million Central Europeans (including many Jews) in the period 1880–1910. In all, some 34 million people emigrated from Europe to the United States in the period 1840–1960.[35]

Upton Sinclair's *The Jungle* (1906) provides a vivid picture of the difficulties endured by the many waves of immigrants in the Chicago meatpacking industry. But America has thrived on the backs of such immigrants, as well as on its black population. For the most part, the offspring of immigrants have multiplied and flourished. The majority of the immigrants to America were anxious to learn English as quickly as possible. On an even larger scale than occurred in Britain after the Education Act of 1870, Robert McCrum, William Cran, and Robert MacNeil argue, "in the schoolroom, and especially the playground, there were fierce pressures favoring the use of the American standard. The schools were the places where the immigrant children were rapidly Americanized by their playmates."[36]

What the schools achieved in relation to Standard American English, the railways supplemented by facilitating the intermixture of a nation that enjoyed an unprecedented degree of mobility from coast to coast. Just as with England's earlier English, American English demonstrated a remarkable ability to scavenge words. New words were assimilated not only from the incoming minorities, but also from such sources as the language of the railways, the paddle steamers, the gangsters, the cardsharpers, the goldfields, and the cowboys. In addition, the Americans have taken over from England the responsibility for providing the cutting edge of English. The old British injunction against parking that involved a full paragraph of words has become "tow away." With this go "fast food" and "take-outs," or "takeovers" and "sellouts." Good innovations in language are often understood without explanation, and in such innovations America has taken the lead. Noah Webster's assumption that British and American English were languages following divergent paths has, however, proved to be far from correct. Quite the contrary. Since Webster's time, both forms of English have become more and more rationalized. Like the English at the BBC, American television newscasters have developed what is known as "Network Standard." One can, of course, exaggerate the peripheral differences, but the fact is that ordinary speakers of American and British Standard English understand each other very well indeed. If that were not so, current societal pressures would force modifications. The case for Standard English in the United States, England, and the other English-speaking countries of the Commonwealth does not have to be made. It is self-evident.

What still remains in doubt is the future of English in the remainder of the British Commonwealth and in the countries outside. India provides the leading example of the other countries in what was once the British Commonwealth. Its population not only dwarfs all the rest, but it also has no particular allegiance to either England or the United States. In the 1950 constitution, Hindi was to be the first national language among the fourteen recognized languages of India. According to Prime Minister Nehru, India was to be shaken free of English by 1965. In fact, quite the reverse has taken place. After the language riots in the South, in May 1963, against the domination of Hindi, a "three-language formula" for official documents and parliamentary business was eventually established. The "formula" included English, Hindi, and one other language.

In India it had taken three languages (instead of two in the Canadian formula) to pacify ruffled feelings. Such feelings result from the de facto predominance of English among the ruling classes and the upwardly mobile. Of more than 750 million Indians, some 7 to 9 percent use English as either their first or second language. The figure is higher than the number who speak English in England itself. In 1978, out of nearly sixteen thousand newspapers registered in India, about three thousand were in English. Had Prime Minister Nehru spent over two years in a small ship traveling round the coasts of British India and Burma, as I did during World War II, he might not have been quite so blinded by his understandable nationalism. The fact is that English was then, as it is even more now, the de facto language that unites very different people. After Indira Gandhi was assassinated, her son Rajiv Gandhi spoke to his people on television in English when he wanted to stop the violence that followed. It is, of course, true that Indians will make English their own, by giving it some coloration from their Asian experience, but they will not want their language to differ radically from Standard English.

As it is with India, so it will be with Pakistan, Bangladesh, Sri Lanka, Malaysia, South Africa, Nigeria, Singapore, Hong Kong, and so many others. But eventually people in these nations—those who need to employ English for international communication in politics, commerce, or science—will use their influence in society to force the development of a more standard language. After all, English is the language of 80 percent of international scientific papers and of information stored in the world's computers. Even in Europe nearly half of the business deals are conducted in English.[37] This does not mean, nor should it, that the English of Third World countries will lose all of its individual colorations. In fact, the economic developments in the Third World strongly suggest that such countries will be contributing to the future evolution of English as a whole. They will do this much as the United States has been contributing to English throughout the present century.

Those who predict that English, like the vulgar Latin, is about to break up into mutually unintelligible languages ignore the effects of the global village, information technology, science, television, films, radio, records, telephones, and the concentrated yearning of 5 billion people for the products of Western technology. (Television alone—with its 750 million TV sets and 2.5 billion viewers—is actively shrinking the size of the globe by bringing news, sports, Western standards, and Western advertising into the living rooms of the world.) It is, of course, true that in the West Indies, as elsewhere, there are strong movements toward forms of English that would be almost unintelligible for outsiders. But the professional classes in such areas are already resisting these moves. Those who speak the more deviant dialects will be obliged to follow the example of Hardy's Tess and modify their accents if they wish to work in the professional or international arenas.

There is nothing new in the idea that people who speak in a dialect are usually also able to speak in the standard language, when this is necessary. Forty years ago, when I spoke to the rural natives of the island of Jersey, we would converse in what they called "Bon Français," rather than in their local patois of "Jerriais." In the German cantons of Switzerland, we would likewise speak in High German rather than in Swiss German. Today, many of the children of such people are accustomed to using English, if only as a second language, and their own patois is dying out. Two hundred years ago, Robert Burns, the Romantic, wrote his finest poetry in a dialect that was almost unintelligible to the Sassenach. But this should not blind us to the fact that his command of English was by no means limited to the dialect. Which brings us back to Hardy's Tess, in 1891, speaking in the dialect when at home and in Standard English when outside her local environment.

For those who learn English as a second language, in all the countries outside the United States and the British Commonwealth, the question of whether to learn Standard English does not arise. Though they sometimes achieve it, they do not generally intend to seek a local coloration for their English. If at all possible, they will seek out, as appropriate, teachers from England or the United States. At the very least, they will listen to English spoken with standard accents, frequently on BBC or Berlitz tapes. More than one-fifth of all the people on earth either speak English or use it as a second language. These people are, for the most part, either those who enjoy the fruits of Western technology or those who yearn for them.

In Japan it is estimated that more than twenty thousand foreign words, mainly English, have entered the vocabulary since World War II. When one visits Japan, it is quickly evident that almost everybody under thirty-five is anxious for the opportunity to speak English with a foreigner. More recently, even the one billion Chinese have quietly dropped the language of their Russian comrades and have taken up English as their second language. The bulk of international politics, business, finance, air control, and much else is now carried on in English. It is therefore natural that, for those who are not born to it, English has become by far the most important second language in the world. By very conservative estimates, three or four hundred million people already use English as a second language.[38] Swift's brilliant but indirect compliment to the Right Honourable Lord John Sommers, in *Tale of a Tub*, goes as follows: "I have somewhere heard it as a Maxim, that those to whom every Body allows the second Place, have an undoubted Title to the First." Swift would be delighted to learn that, on the basis of these criteria, English has by now easily outdistanced French, which in his time had been its chief rival.

III Rationalizing Production

VII Great Britain and the Industrial Revolution

Work study, or industrial engineering, divides naturally into time measurement and method study. Time or work measurement indicates how long a particular element of work may be expected to take. Method study (which is closely related) determines where and when—as well as by which person or machine—such elements of work can best be undertaken. This form of study (which has gone by far too many names, and sometimes none at all) has spearheaded the rationalization of manufacture. It is fundamental to Western production methods and, like so much with which we have been concerned, is inextricably bound up with the measurement of time.

In the second part of the seventeenth century, a number of conditions contributing to industrialization made England particularly suited for the early development of time and motion study.[1] These factors included the trebling of London's population during the century; the benefits to shipping of the (anti-Dutch) Navigation Act of 1651; the rebuilding of London after the Great Plague of 1665 and the Great Fire of 1666; the rise of the "Puritan" ethic; the use of the division of labor; the birth of modern science, symbolized by the founding of the Royal Society in 1662; the development of financial institutions, symbolized by the founding of the Bank of England in 1694; and the beginning of modern mechanical technology, symbolized by the use of the pendulum in clocks from 1657. In addition, England was at this period particularly receptive to competent men and profitable ideas. As John Dryden put it: "The genius of our countreymen, in general, being rather to improve an invention than invent themselves . . . is evident not only in our poetry but in many of our manufactures." Defoe says "that there never was known such a trade all over England as was in the first seven years after the plague and after the fire of London."[2]

As in West Germany and Japan after World War II, the plague and fire in Restoration London provided a great impetus for production. The elements that we have mentioned are part of the equation, and they are impressive. They have given us what all the world desires: Western technology. But a prerequisite for the rationalization of production is a prior rationalization of all those areas of measurement and communication with which we have thus far been concerned. Without a sense and measure of time, we cannot introduce time studies; without reliable weights and measures, we cannot rationalize batch production or introduce interchangeable parts; and without rationalized language and numbers, we can neither document our methods nor record our results.

Let us remind ourselves of the earlier lack of rationalization that pertained to agricultural and related work methods. Walter de Henley (ca. 1240) wrote a book on husbandry for the benefit of his son. He is aware that a ploughman should be able to work a given number of acres per day with oxen or horses, according to the nature of the land.[3] But since the length of his day is related to the time of the year (and even the size of an acre varied), the study, though valuable, is clearly very much circumscribed. Even when Leonardo da Vinci (1452–1519) carried out a much more detailed time study on the shoveling of earth, the "equal" hour was not universally accepted, and "the 'foot,' the piede, had a value of 17.134 inches in Milan, 14.07 inches in Padua, but 11.73 inches in Rome." Much the same is true of the "braccio," or "arm," used in the study; the braccio is a "length often quoted in Leonardo's notes."[4]

The sandglass or the water-clock (clepsydra) could be employed for timing a simple operation. Among the Science Museum's display of sandglasses, in London, is the illustration of a medieval stamping mill from *Das Feuerwerkpuch*, a treatise on the manufacture of fireworks written in Germany about 1450. An operation is clearly being timed by the sandglass shown on the left (figure 7). In a comparable way, simple water clocks were used in North Africa to denote the period of time during which landowners were entitled to extract water for irrigation.

Figure 7. Part of an illustration from *Das Feuerwerkpuch*, ca. 1450. (The Board of Trustees of the Royal Armouries, London)

We also know from Lucian's *Fisherman* that, under Roman rule, the length of time for which a lawyer might plead a case was customarily controlled by clepsydrae.[5]

However, many prerequisites for modern production methods began in the seventeenth century. The Cartesian analogy of man (and dog) with a machine is a product of that period. It was reflected not only in the mechanical theory of John Locke's association of ideas, but, also more directly, through David Hartley's *Observations on Man* (1749). In the modern era we find these ideas reflected in the behavioral sciences (such as industrial psychology and cybernetics), which are very much the present and future concern of work study. That men can be characterized as wound up, run down, rusty, or going like clockwork is essentially a product of seventeenth-century thought. It implies not only that their work may be accurately measured, but also that their motions can be studied. Given the proper incentives, they will follow predetermined and appropriately mechanical patterns. But motion study (an element of method study) requires also the ability to describe and to portray graphically the nature of trades and the movements of men, tools, machines, and products. And in this respect, too, the late seventeenth and early eighteenth centuries were a critical period. The movement to describe trades is demonstrated in particular by Chambers's *Cyclopaedia* (1728), Diderot's *Encyclopédie* (1751–65), and the *Encyclopedia Britannica*, which was issued from 1768 as a "Dictionary of Arts and Sciences."

THE DIVISION OF LABOR AND OTHER DEVELOPMENTS IN INDUSTRY

A further precondition for productivity science, one that is related to both time and motion (or method) study, is the division of labor. The division of human labor into trades was an early concomitant of urbanization. In Plato's *Republic* the case for specialized administrative functions is consistently argued through an analogy with the trades: "More things of each kind are produced, and better and easier, when one man works at one thing, which suits his nature, and at the proper time, and leaves the others alone."[6] But subdividing the trades themselves—by consciously adopting the process known as "division of labor"—was a further rationalization that did not become established practice until relatively modern times. In general, it presupposes the weakening of the guilds and the demise of the master craftsman, who was responsible for making a total product.

In this respect there is a decisive difference between the description of trades by pioneers—such as Boyle or the anonymous author of *Humane Industry; or, A History of Most Manual Arts* (1661) in the middle of the seventeenth century—and the *Encyclopédie* in the middle of the eighteenth century. In the latter work the procedures are not only graphically portrayed, but the labor is frequently divided into a large number of processes in each of which workers specialize. The *Encyclopédie* divides clockmaking into sixteen and watchmaking into twenty-one processes that are described clearly and in detail. The point is then made that each part of a clock or watch must be perfect because it is produced by a worker who specializes in a single component.[7] This type of carefully documented analysis is one of the first steps in productivity science, and the authors of the *Encyclopédie* are conscious of the importance of their work for the development of industry.

As with so many of the preconditions for time and motion study, horology also contributed significantly to the division of labor. Thomas Tompion, the father of English clockmaking, had produced six thousand watches and five hundred clocks by the time he died in 1713. The unprecedented demand appears to have obliged him to initiate a form of batch production, a first step toward the division of labor. Sir William Petty, the economist, is probably referring to Tompion's workshop when he argues for the advantages of the division of labor: "As for Example, in the making of a *Watch*, if one man should make the *Wheels*, another the *Spring*, another shall Engrave the *Dial-Plate* and another shall make the *Cases*, then the *Watch* will be better and cheaper, than if the whole work be put upon any one Man."[8] The topical nature of this subject in the first half of the eighteenth century is further demonstrated by Bernard Mandeville, who uses the success of clock manufacture as an argument for the division of labor, in his *Fable of the Bees*. Alexander Pope humorously suggests that bad poets might emulate the methods of clockmakers and "manufacture" poetry, each using what little specialization he has. But the way that he introduces his facetious argument, in *Peri Bathous* (1728), is particularly enlightening for our purpose: "The vast improvement of modern manufactures ariseth from their being divided into several branches."[9]

Fifty years later, in 1776, Adam Smith singled out watch movements as those articles whose price had been most reduced (from twenty pounds to twenty shillings), despite the inflation of the previous hundred years.[10] When Charles Babbage wrote his *Economy of Machinery and Manufactures* (1832), he reported that watchmaking in his time was divided into no less than 102 different trades. It would therefore not surprise work-study engineers to learn that the first really detailed study dealt with what appeared to be a more simple product than watches.

M. Perronet's *Art de l'Epingler* (1762) provides a glossary of terms for pinmaking. He also describes individually the 130 illustrations on seven folio sheets, through which the pinmaking trade is portrayed. The study itself includes detailed statistics for both the timing and the costing of each operation.[11] It is sufficiently specific for Charles Babbage, some seventy years later, to make direct comparisons with his own figures for pinmaking. Perronet's study should be seen as part of an ongoing interest in the division of labor during the eighteenth century. He divides pinmaking into sixteen processes, Ephraim Chambers's *Cyclopaedia; or, An Universal Dictionary of Arts and Sciences* (1728) lists twenty-five operations, and the *Encyclopédie* has eighteen. Adam Smith not only begins his *Wealth of Nations* (1776) by stressing the importance of the division of labor, but he uses as his first example the "about eighteen operations" in pinmaking.

Clearly, the division of labor entered the industrial "climate of opinion" from the late seventeenth century. With the wisdom of hindsight, we can wonder why trades with such a potential turnover as coining, thimblemaking, and pinmaking—the *Münzmeister, Fingerhüter*, and *Nadler* are illustrated in Amman's *Stände und Handwerker* (1568)— were not rationalized earlier. But first, new pressures had to break down established methods of manufacture, particularly those of the trade guilds, and provide the underpinning for more modern methods. In England, some factors that provided this impetus were the increased mobility of labor, the improvement in education, the increasing prestige of science, the tendency to upgrade "mechanical arts" (particularly horology and instrument making), the growth of financial institutions, the accumulation of capital, the improvement of transport (the turnpiking of main roads was virtually completed when canal construction began in the 1750s), and the potential for standardization that resulted from an increasing market at home and abroad. An important feature of the eighteenth century was the proliferation and growth of towns in northern England, related in part to increased agricultural productivity and further enclosure of common lands.

If we return to pinmaking for a moment, we may note that the developments discussed thus far were part of an ongoing process. Henry Hamilton tells us that pin manufacture in England was "probably of little importance before the middle of the sixteenth century." Yet half a century later in a petition to James I the pinmakers "declared that there were no less than 20,000 people . . . employed in making pins in England." By 1697 the consumer demand that this implies had produced "a classic illustration of division of labour." At Dockwra's Copper Company, specializing in pins, "it was said that some 'top workmen' could deal with 24,000 a day."[12]

Until about 1700 battery and wire drawing (as used in the pin trade) were the two main processes in the brass and copper industry. After this, casting (which had earlier been used for bells and cannon) was introduced for the production of small household articles. Casting implied a standardization of the product. The first account of such a process that we have is for thimbles cast about six gross at a time in Highgate, during the last decade of the seventeenth century. From about 1700, rolling mills began to replace the old process of battery, and paved the way for Birmingham's stamped brass-foundry trade. This further revolutionized the manufacture of a very wide range of standardized household articles.[13]

In other areas of manufacture, we learn of nine frying-pan plates being battered at one time "like a nest of Crucibles or Boxes" as early as 1686, and in 1705 a letter from a grinder of convex glasses for telescopes complains about a special order because the customer is not yet aware that "the new way of making them is by working 4, 6, or 8 together."[14] In coinage, which was produced by battery until 1662, demand overtook supply to such an extent that by 1753, despite the use of the fly press, half the copper coinage was estimated to be counterfeit. By the end of 1788 Matthew Boulton had to set up new presses in a manner intended to cut down to a minimum the expense and potential errors of human handling. By 1790 he had patented a coining press to be driven by steam "in place of men's labour, as has hitherto been practised."[15] One senses in such developments the beginnings of standardized products and standardized production methods.

TIME STUDY, PIECE WORK, AND ORGANIZATION

Ultimately, the rationalization involved in the division of labor led to mass production through mechanization, but more immediately it lent itself to time study and piece work. With respect to time study, the first stop watch was Samuel Watson's "pulse watch," invented for use by physicians about 1690.[16] By 1761 stop watches of a more advanced design had achieved enough popularity for Laurence Sterne to be able to refer to their use at some length in *Tristram Shandy* (3.12). The first time recorders for clocking workmen into the factory belong to the same period; they were invented by John Whitehurst of Derby about 1750. There was one in the Etruria factory of his friend Josiah Wedgwood, who was particularly time conscious. R. M. Currie quotes a document of 1792 in which a certain Thomas Mason promises to use his "utmost caution at all times" to prevent anyone knowing that he is "employed to use a stop watch to make observations of work done" at the Old Derby China Factory.[17]

This is not an isolated occurrence. While the work of Boulton and Watt in manufacturing steam engines is deservedly well known, the achievement of their respective sons—who greatly improved productivity through time and work-flow studies after the lapse of the patent on the steam engine in 1800—has received far too little attention. The new construction in the Soho factory under the particular stimulus of James Watt, Jr., took place during 1795–1801, and provides "the earliest records in existence of systematic factory planning." The ten shops designated A–K were placed in logical sequence to reduce handling; standard and generally increased speeds (in the future tradition of Frederick Winslow Taylor) were set for machines; and a "specification" of 1801 "routes" the sequence of shops in which itemized work is to be undertaken. But James Watt, Jr., went beyond the application of the division of labor to factory routine. He was involved in detailed cost accounting which allowed him to list the profit and loss in each of his ten shops. Itemized time studies were also made for boring, turning, and fitting over a range of sizes.[18]

Even in the case of turning, where piece rates might have been set with little difficulty, Watt appears to have preferred paying a "gratuity" of one-sixth of the price at which each job was costed, based on the time that it would take. It is clear from the balance sheets that the new methods paid handsomely despite and perhaps even because of the loss of the firm's monopoly after 1800: when James Watt, Jr.—"the pioneer of time-and-motion study and engineering costing"—died in 1848, he left a fortune of £160,000.[19] Watt may have been the pioneer, but he was not alone. In 1839 a description of Sir William Fairbairn's Manchester works for steam and locomotive engines says: "There are from 550 to 600 hands employed. . . . In every direction the utmost *system* prevails, and each mechanic appears to have his peculiar description of work assigned with the utmost economical subdivision of labour."[20]

In the early nineteenth century, certain mechanical tools (like textile machines) had been mass produced for some years, but the "idea of having ready-made machine-tools was a novel one." As A. E. Musson and Eric Robinson put it, James Nasmyth found himself with so considerable a demand that in his Bridgewater Foundry (completed 1837) he "appears to have been one of the pioneers of assembly-line production," and "clearly recognized the advantages of line-production and a smooth flow of work."[21] It must be stressed that factory organization is not an exclusive attribute of the twentieth century. Indeed, at the very beginning of the period with which we are concerned, Ambrose Crowley III brought hand, rather than machine, workers together in an early form of factory organization that would employ a thousand men and continue into the nineteenth century. He divided his anchor makers into gangs specializing in particular sizes, and organized at Winlaton a large number of separate workshops. Each specialized in making particular nails (Crowley's main product, with over a hundred varieties), or in such other activities as frying-pan, hoe, and file making.[22]

M. W. Flinn notes that since Ambrose III insisted on spending most of his time at the sales outlets in London and Greenwich—some three hundred miles from his works council—he set up his own system of time measurement: "Because accounts were submitted weekly from the factories in the North . . . the weeks were referred to as *accounts*. 'Account I' may be traced back to November 1685 . . . [and] by 1816 Account 6808 had been reached!" Ambrose III organized piece-work payments, and in 1690 he set up a "Court of Arbitrators," in which two of the five arbitrators were elected by employees.[23]

Apart from the complex organization involved in running his factories, Ambrose III provided a wide range of social services more than a century before Robert Owen, who is generally considered the first factory owner with a modern social conscience. In connection with a fund to which the employer as well as the employees contributed, Ambrose III instituted education and social insurance for his workers. He also provided free medical care, medical supplies, a chaplain with duties related to social work, and cheap housing. Sir Eric Roll—referring to the introduction of an insurance society at the Soho works of Boulton and Paul, at some time before 1792—suggests a range of selfish and paternalistic motives, such as attaching the hands to the factory, saving on the poor rates, and increasing one's power over the workers.[24] Much of this is probably true, but it is also worth noting that Crowley's action coincides with the beginnings of the sentimental movement in literature. The bourgeois "benevolence" that this movement reflects produced not only tear-jerking plays and novels, but also an attitude toward the unfortunate that was generally less inconsiderate than what Allardyce Nicoll calls the "cynical aristocratic existence of former times."[25]

MAN AGAINST MACHINE

One of the paradoxes of the British industrial revolution does not appear to be repeating itself in our incipient global industrial revolution. In England a remarkable reduction in the price of both products and raw materials (deflation!) coincided with a parallel reduction in the living and working standards of factory hands, particularly women and children. M. W. Flinn demonstrates the increase in purchasing power:

> The inventions in cotton spinning resulted in the successive reduction in the price of cotton yarn from 38s per lb in 1786 to 2s 11d in 1832; Darby's and Cort's inventions, by freeing the ironmaster from the threat of rising marginal costs of charcoal, and by raising the productivity of both labour and capital in the process, brought down the price of a ton of bar iron from about 18 pounds per ton in the middle of the century to about 8 pounds in the early 1820s; and the Duke of Bridgwater's first canal reduced the delivered price of coal

in Manchester overnight from eightpence a hundredweight to fourpence.

As a result, "real purchasing power was increased." Among a wide range of examples, Babbage points to a brass knob made in Birmingham twenty years earlier at thirteen shillings and fourpence per dozen. It was now being turned by a steam engine and the superior product sold at one shilling and ninepence farthing per dozen, less than one-seventh of the former price.[26]

Part of the pressure being placed on factory hands resulted from competition with power machinery, though we have noted from the experience of Boulton and Watt—as well as Fairbairn and Nasmyth—that successful factory owners had to do more than merely install machines. Nor, on the other hand, could those who did not introduce machinery rely only on reducing piecework rates and increasing work hours. The more enterprising manufacturers, as we have seen, were beginning to initiate time and method study.

The beginning of the trend to increase the pressure on the weakest elements in labor is reflected in literature by Goldsmith's *Deserted Village* (1770) and Blake's *Songs of Experience* (1794). It becomes more clear in Robert Owen's *Observations on the Effect of the Manufacturing System* (1815). Owen tells us that "not more than 30 years since, the poorest parents thought the age of 14 sufficiently early for their children to commence regular labour." But now, "in the manufacturing districts it is common for parents to send their children of both sexes at seven or eight years of age, in winter as well as summer, at six o'clock in the morning." After describing the highly unsuitable working conditions, Owen points out that the children "remain, in a majority of cases, till eight o'clock at night." Though Owen is a remarkably enlightened employer for his time, his argument for reducing working hours (and increasing the age at which boys and girls begin to work) foreshadows Charles Eugene Bedaux's relaxation and fatigue allowances, even to the extent of being promoted on the basis of expediency. Owen says, in part, that workers employed for twelve hours daily will produce virtually as much as those factories "in which the exertions of the employed are continued to 14 or 15 hours per day."[27]

The developing juxtaposition of handworker and power machine was a contemporary phenomenon for Charles Babbage, whose *Economy of*

Machinery and Manufactures (1832) is still a valuable and fascinating book for anyone interested in productivity. In addressing the question of man versus machine at the conclusion of his chapter on "Proper Circumstances for the Application of Machinery," Babbage begins: "One of the most common effects of the introduction of new machinery into manufactures, is to drive out of employment much of the hand-labour which was previously used." He notes that whereas there had been 240,000 hand looms in England and Wales in both 1820 and 1830, the number of power looms, each doing three times as much work, had increased during that time from 14,000 to 55,000, causing "considerable suffering amongst the working classes."

Babbage—who was (like Newton before him) the Lucasian professor of mathematics at Cambridge—then addresses himself to a public relations problem some of whose aspects are not unknown to work-study practitioners in our own time: Is it in the interests of the working classes that improved machinery should be so perfect as to defy the competition of hand labor, or is it better for them to be forced to quit the trade gradually *"by the slow and successive advances of the machine"*? Babbage considers that a quick transition causes less permanent damage and encourages the workman "to learn a new department of his art." He feels that "it is almost the invariable consequence of such improvements ultimately to cause a greater demand for labour," but that workpeople are entitled to be informed of the facts "at an early period, in order to diminish, as much as possible, their injurious results." As a personal insurance against industrial change, he recommends diversity of employment within the family, and membership in Friendly Societies and savings banks.[28]

BABBAGE ON TIME STUDY, METHOD STUDY, AND THE "DIVISION OF MENTAL LABOUR"

In R. M. Currie's *Work Study*, the first paragraph of the chapter "Method Study: Install and Maintain," mentions the need for "finding alternative work for any workers who may be displaced as a result of the new method" (something that is seemingly of more concern to the Japanese nowadays than it is to us). Babbage was equally involved with the public relations aspects of "installing the new method" one and a half centuries ago. In his chapter "On the Method of Observing Manufactories," he first sets out a standard form and suggests that it be printed. This form provides questions regarding the nature and processes of the particular trade being studied. Wherever possible, numerical answers are preferred. A subsequent form questions the nature of each particular process, including the number of times "the operation is repeated per day or per hour." Babbage is keenly aware of certain human problems related to time study that have hardly changed:

> In filling up the answers which require numbers, some care should be taken: for instance, if the observer stands with the watch in his hand before a person heading a pin, the workman will almost certainly increase his speed, and the estimate will be too large. A much better average will result from enquiring what quantity is considered a fair day's work. When this cannot be ascertained, the number of operations performed in a given time may frequently be counted when the workman is quite unconscious that any person is observing him. Thus the sound made by the motion of a loom may enable the observer to count the number of strokes per minute, even though he is outside the building in which it is contained.[29]

Babbage here uses the term "a fair day's work" long before it became institutionalized by Frederick Winslow Taylor. But even Babbage is very conscious that he is discussing an established procedure. He quotes the published advice of M. Coulomb ("who had great experience in making such observations") to the effect that work should be timed at different periods of the day without the workers knowing that they are being observed. One wonders why, in 1792, Thomas Mason, at the Old Derby China Factory, had been required to pledge solemnly that he would "prevent the knowledge transpiring" that he was "employed to use a stopwatch to make observation of work done." Perhaps the proprietor's order of priority was first to avoid a change in the worker's "rating," second to retain secrecy with respect to competitors, and only last of all to consider the rights and feelings of the worker.

In his chapter "On the Division of Labour," Babbage first lists a number of the advantages of this procedure proposed by other writers: less time is taken in learning, and less material is wasted; less time is lost in changing mind, hand, and tools from one process to another; and a particular skill is acquired by frequent repetition of the same process. In addition, specialization is more likely to suggest improvements in the form of tools and the mode of using them, as well as providing "the first steps towards a machine." Babbage recognizes that the previous lack of "an extensive knowledge of machinery, and the power of making mechanical drawings" was "one of the causes of the multitude of failures in the early history of many of our manufactures."

Thus far, Babbage has gone little further than others like Adam Smith, but he then turns to "what appears . . . to have been omitted by those who previously treated the subject." He quotes from Perronet's studies of pinmaking, and compares them with extensive time and cost studies of his own. The crux of his argument is that, over and above the advantages previously enumerated, the "master manufacturer . . . can purchase exactly" that precise mix of labor which will provide the most effective and economical production unit. Thus, a skilled man earning six shillings per day for whitening the pins will not be spending part of his time putting on the heads at one shilling and sixpence for every twenty thousand; that was an occupation at which a woman might expect to earn one shilling and threepence per day or a child even less. Babbage is proud of his contribution to a procedure that is part of what we now call method study ("This principle presented itself to me after a personal examination of a number of manufactories and workshops devoted to different purposes"), but his academic integrity obliges him to add that the principle had been independently enunciated by Gioja in a publication of 1815. Clearly—as we have seen with discoveries as different as the calculus, evolution, and atomic fission—the same perceptions were often made independently, in part because the time was opportune for them.

Babbage stresses that his contribution to method study would have been shown to provide even more remarkable savings had he selected as his illustration the art of needlemaking, in which the difference between the greatest and the least earnings were forty-fold. But as a true exponent of work study he does not automatically assume that a machine will produce more effectively than a man in all situations. He also discusses a machine for making pins, recently discovered by an American. Here—as even more specifically when later considering the new bobbin-net machines—he has a keen sense not only of time and method, but also of the capital costs and depreciation involved in installing new equipment. This is a study that companies which, more recently, were overhasty to install expensive computers would have benefited from reading.

Babbage's chapter "On the Division of Labour" is followed by a chapter entitled "On the Division of Mental Labour." He applies to mental activities his principle that labor should be employed only in the work for which it has been specifically trained. His main example is a project for preparing extensive mathematical tables by using three groups of people: a small committee of leading mathematicians to provide the relevant formulae; a section of seven or eight lesser mathematicians to "convert into numbers the formulae put into their hands"; and a larger section of sixty to eighty persons to carry out the "mechanical" calculations.

In the fourth edition of *Economy of Machinery and Manufactures*, Babbage, the inventor of the computer, announced that "since the publication of the Second Edition of this work, one portion of the [calculating] engine which I have been constructing for some years past has been put together." In the later editions, he also added to this chapter a paragraph listing ten separate managerial categories and pointed out the great improvements to mining that had resulted from the "judicious distribution" of management duties "which have gradually been introduced."

THE RELATIONSHIP BETWEEN WORK STUDY AND MANAGEMENT

The early manufacturers—one has only to think of Crowley, Boulton, Wedgwood, and Arkwright—undertook much of their own management. They were usually self-made men who frequently had to train themselves, their "hands," and even their foremen. There were few enough patterns that they could follow, no schools of management on which they could draw, and no studies by Babbage, Taylor, or Currie setting out procedures. What records we do have of the organization and methods that they and others developed come down to us more by chance than by intention. They were certainly not published for the benefit of their competitors. This may help to suggest why so little is known about the early history of work study, though it had certainly started by the end of the seventeenth century, and many of its modern functions existed at least in embryo by the time of Babbage.

James Watt, Jr., now rightly regarded as a pioneer of time and motion study, is also an important transitional figure when one considers impending changes in the effective control of industry. He first entered the Soho Works in what was virtually a managerial capacity for his father and his father's partner. Faced with potentially disastrous competition when the "monopoly" on steam engines ended in 1800, he developed time and method study to such an extent that the works did far better under the second generation than they had done under the first. But still being essentially an entrepreneur rather than an academic or a professional manager, he had no interest in broadcasting the "mystery" of his achievement. A comparable mystique surrounded later consultants like Bedaux, but by his time the nature of business had changed and the ownership of capital was becoming more and more divorced from management. In the United States, after 1926, Charles Eugene Bedaux found himself employed by large corporations for whom he was obliged to train time-and-motion-study engineers. This meant that his ideas could not be kept secret indefinitely. Indeed, I myself had the good fortune to be instructed by one of his trainees.

The fact that managers are now employees rather than entrepreneurs has contributed to professionalism in management and to increasingly standardized methods for educating managers. But management has become more than a single discrete entity; it seems to be evolving along lines analogous to the division of labor, because there has been increasing specialization within management itself. The trend was foreshadowed by the innovative model of ten managerial functions for mining, which Babbage added to his chapter "On the Division of Mental Labour." By the end of the nineteenth century Taylor would also recognize the need for specialization in management. This is reflected in particular in his concept of the "functional or divided foremanship." Functional foremanship "consists in so dividing the work of management that each man from the assistant superintendant down shall have as few functions as possible to perform."[30] There are few of us today who have not been affected in one way or another by Taylor's formula for the functions of management.

VIII North America and the World

It is with Frederick Winslow Taylor (1856–1915), born in Germantown, Pennsylvania, that work study or industrial engineering has normally been thought to begin.[1] Since Taylor's work was concerned essentially though not exclusively with time study, it is a commonplace to credit him, as does Currie, with "having first evolved the principle of breaking a job down into detailed elements to determine a time to be allowed for the job." We have seen, however, that time studies, like so much else, really began in the Restoration and eighteenth century. Taylor was neither a factory owner like James Watt, Jr., nor a scholar like Babbage, but an employee who had been apprenticed as a pattern-maker and machinist. As early as 1880 or 1881, when he was gang boss at the Midvale Steel Company, Taylor realized the need for measuring a "fair day's work" on any operation. This led to a series of carefully planned investigations by himself and other like-minded men. They included Henry L. Gantt, his assistant at Midvale, and F. A. Halsey, who, like Gantt, is known for his studies on incentives. Taylor preferred to call his work "Scientific Management" and disliked the term "Taylor System."[2]

In the Spanish-American War, Taylor, who had joined Bethlehem Steel in 1898, had the opportunity of demonstrating how eighty thousand tons of pig iron, unsaleable before the war, could now be loaded in just over one-quarter of the man-hours that were normally required (*Principles*, p. 41; *Shop Management*, pp. 49–50). As a result of work measurement, Taylor estimated that certain selected men with qualities "of the type of the ox" could move forty-seven tons per day instead of the normal maximum of twelve and a half tons. He first hired a specially selected workman and offered him a consistent payment of $1.85 per day (instead of $1.15 per day). Taylor knew, through earlier work done with Carl G. Barth, that there was a direct relationship between the time that a man could be under load and the weight being carried. (In this case, a forty-two-pound pig of iron could only be carried for 43 percent of the working day.) He promised the extra wages on the understanding that the workman, whose name was Schmidt, agreed to "having a man . . . who understood this law, stand over him and direct this work day after day, until he acquired the habit of resting at proper intervals" (*Principles*, p. 59).

It will be noted that Taylor paid to the workman only a small proportion of what he saved for management and, in his view, ultimately for the public. In *Shop Management* Taylor defends the Towne-Halsey system of limited incentives by saying: "This system . . . diminishes soldiering, and . . . since the workman only receives say one-third of the increase in pay that he would get under corresponding conditions on piece work, there is not the same temptation for the employer to cut prices" (*Shop Management*, p. 39). Taylor has a good deal to say about "soldiering," or shirking work (which he tells us is called "hanging it out" in England and "ca' cannie" in Scotland). He points out, for example, that when he arrived at the Midvale Steel Company in 1878, he soon realized that the workmen "had set a pace for each machine throughout the shop, which was limited to about a third of a good day's work" (*Principles*, p. 49). Taylor sympathizes with the natural fears and superficial prudence which would lead men to take such action, but he feels strongly that it is not in the interests of the workers, the employers, or the consuming public that this should occur. He points out that the workmen in "England, more than in any other civilized country, are deliberately restricting their output because they are possessed by the fallacy that it is against their best interest for each man to work as hard as he can" (*Principles*, p. 142).

Taylor believed that both management and labor had to be restrained from taking advantage of "the third party (the whole people)." Giving labor an increase of much more than 60 percent in wages, despite a far greater increase in efficiency, would mean that "many of them will work irregularly and tend to become more or less shiftless, extravagant and dissipated" (*Principles*, p. 74). But he also felt strongly that the public "will no longer tolerate the type of employer who has his eye on dividends alone, who refuses to do his full share of the work, and who merely cracks his whip over the heads of his workmen and attempts to drive them into harder work for low pay." Nevertheless, Taylor felt, too, that the public would "no more . . . tolerate tyranny on the part of labor which demands one increase after another in pay and shorter hours while at the same time it becomes less instead of more efficient" (*Principles*, p. 139). This, perhaps even more than the employers who crack whips, remains a problem that is still very far from having been resolved.

With shoveling, as with the loading of pig iron, Taylor feels that "the average man would question whether there is much of any science in the work" (*Principles*, p. 64). Yet by a series of carefully controlled studies—which remind one of the earlier and less precisely documented work of Leonardo da Vinci before the rationalization of measurements—Taylor demonstrated that a man did more work per day with a shovel load averaging twenty-one pounds, "whatever the class of material they were to handle," than with any other size (*Principles*, p. 66; *Testimony*, pp. 54–55). In the classic series of experiments related to this work, Taylor produced "8–10 different kinds of shovels . . . appropriate to handling a given type of material" (*Principles*, p. 66); he broke up the work gangs; and, at the beginning of the working day, assigned specific tasks to each man and provided him with a record of his success in the previous day's task. The result was a reduction of work force from between 400 and 600 to 140 men; an increase in moving sixteen tons per man-day for $1.15 to moving fifty-nine tons per man-day for $1.85; and a decrease in the cost of shoveling one ton from 72 cents to 33 cents, despite the very considerable increase in managerial and clerical work that Taylor's methods involved.[3]

But Taylor was very much aware of the working man's fears of being put out of a job through the effectiveness of Scientific Management. He used as an example the fear that existed among English hand weavers when it appeared that the power loom would put them out of work: "In Manchester, England, in 1840, there were 5,000 operatives, and in Manchester today there are 250,000 operatives . . . there now comes out of Manchester, England, 400–500 yards of cotton cloth for every single yard that came out in 1840" (*Testimony*, pp. 16–17).

Behind Taylor's increase in productivity lies the concept of breaking down an operation into simple elements and recording those details: "Thousands of stop-watch observations were made to study just how quickly a laborer . . . can push his shovel into a load of materials and then draw it out properly loaded."[4] Taylor's method was to select and time a worker with above-average potential for the particular job with which he was concerned. He knew that different jobs require different characteristics, and his selection methods seem not only post-Darwinian but even reminiscent of the products of the "bokanovskified egg"—"ninety-six identical twins working ninety-six identical machines"—that Aldous Huxley satirized a generation later in *Brave New World*.[5]

TAYLOR'S EXPERIMENTS WITH INSPECTING BICYCLE BALLS, AND WITH SPEEDS AND FEEDS FOR MACHINE TOOLS

While Taylor sought men "of the type of the ox" for his studies in loading pig iron, his studies on the inspection of bicycle balls involved the selection of girls "born with unusually quick powers of perception accompanied by quick responsive action" (*Principles*, p. 89). At the close of the nineteenth century, America experienced a remarkable boom in the bicycle trade, which then dropped off from 2 million to 250,000 units per year between 1899 and 1904. This might be seen as the temporary eclipse of the mechanical horse before the age of horseless carriages, foreshadowed by Henry Ford's prototype. By the late 1920s, 15 million Model Ts had been sold. In achieving this, Ford used mass production and sales methods analogous to those of the cycle makers. Wilbur and Orville Wright—who made the first powered, sustained, and controlled airplane flights in history, on December 14, 1903—had also learned much from their cycle-repair business.

Writing in *Principles*, Taylor says that "when the bicycle craze was at its height some years ago . . . the writer was given the task of systematizing the largest bicycle ball factory in this country" (*Principles*, p. 86; *Shop Management*, pp. 85ff.). Starting with the 120 girls employed in inspection, he began a process known to modern work-study practitioners as "creaming the business": "In most cases . . . there exist certain imperfections in working conditions which can at once be improved with benefit to all concerned. . . . A most casual study made it evident that a very considerable part of the ten and one half hours during which the girls were supposed to work was really spent in idleness because the working period was too long" (*Principles*, pp. 86–87). When asked to vote on a reduction of work for the same pay, the girls (as occurs with management elsewhere in Taylor's studies) "wanted no innovation of any kind." However, in this study—carried out by Taylor with the personal supervision of Sanford E. Thompson, "perhaps the most experienced man in motion and time study in this country, under the general superintendance of Mr. H. L. Gantt"—the working day was arbitrarily shortened in successive thirty-minute steps from ten and a half to eight and a half hours. Two recess periods of ten minutes each were also introduced in both the morning and afternoon, in order that employees should not work continuously for more than an hour and fifteen minutes. Despite the added clerical and managerial costs, as well as the added cost to quality control of "over-inspecting" the millions of small metal balls for defects, some remarkable savings were involved: "Thirty-five girls did the work formerly done by one hundred and twenty. And . . . the accuracy . . . at the higher speed was two-thirds greater than at the former slow speed." In addition to a shorter working day, employees gained a Saturday half-holiday and an increase in wages that "averaged from 80–100 per cent." As elsewhere, Taylor stresses that "the most friendly relations existed between the management and the employés, which rendered labor troubles of any kind or a strike impossible" (*Principles*, pp. 86–97, 135; *Shop Management*, pp. 85–91). Interestingly, Taylor feels that if a reward is to stimulate productivity it must be given soon after the work has been done. For this reason, "profit sharing" or the sale of stock to employees can be "at the best only mildly effective in stimulating men to work hard" (*Principles*, p. 94).

Taylor's longest, most elusive, and in many ways most satisfying project involved discovering the optimum speeds and feeds when using machine tools. The work commenced in 1881, in the machine shop of the Midvale Steel Company, and continued for some twenty-six years. Taylor estimated that it took thirty to fifty thousand experiments to answer the two basic questions which face "every machinist each time that he does a piece of work in a metal-cutting machine. . . . At what cutting speed shall I run my machine? and what feed shall I use?" (*Principles*, p. 106). Since twelve variables had to be taken into account for setting up any given job, the mathematical problem alone seemed almost insuperable. Among the variables were the quality of the metal to be cut, the chemical composition from which the tool was made, the thickness of the shaving to be removed, the shape of the cutting edge of the tool, the quantity and nature of the cooling medium, and the depth and the duration of the cut. Eventually, Carl G. Barth, with the assistance of H. L. Gantt, "succeeded in developing a slide rule by means of which the entire problem can be accurately and quickly solved by any mechanic" in less than half a minute (*Shop Management*, p. 180; *Principles*, pp. 98–116; *Testimony*, pp. 97–108).

The first variable alone among the twelve to be taken into account when setting up a job—namely, the quality of the metal to be cut—was given one hundred different values by Taylor. As Taylor put the problem: "The metal-cutting machines throughout our machine-shop have practically all been speeded [given optimum speeds] by their makers by guesswork . . . [and] we have found that there is not one machine in a hundred which is speeded by its makers at anywhere near the correct cutting speed" (*Principles*, p. 112). For the purposes of work study, the ability to forecast the time that a machine would take to do a job was itself an invaluable tool in eliminating human inactivity.

Taylor also contributed to the effectiveness of machine tools in quite a different way. Working with J. Mansell-White, he discovered, in 1898, what is now called the Taylor-White process of treating tool steel. For a given example, steel made with this process could cut at the rate of sixty feet per minute against twelve feet per minute with a tool made from the best carbon tool steel (*Shop Management*, p. 124). Taylor's experiments with lathe work produced improved criteria for organization, standardization, and payment for machine production; and his job measurements resulted in payments based on "a proper day's work." Taylor and his associates—whose main work was undertaken during the thirty or so years leading up to World War I—were for the most part Americans. This phenomenon is explained in a lead article in the British journal *Work Study and Management Services*, which says that "the difference between the quality of the two labour forces" is crucial, despite the fact that England was then still "the most highly industrialized country in the world."[6] Two world wars, combined with the disparate attitude toward industrial productivity, would change radically the relative economic positions of Britain and the United States.

BEDAUX: RATING AND RELAXATION ALLOWANCES FOR TIME STUDY

Charles Eugene Bedaux (1887–1944) was a Frenchman who emigrated to the United States at the age of about twenty. He improved on Taylor's ideas about time measurement by "rating" (evaluating) the performance of the worker, and his ideas on rest requirement by incorporating relaxation allowances into time study.

Taylor had measured the actual time taken by a worker in performing a job and used this as the basis for calculating payments. Frequently, as we recall from the pig-iron and bicycle-ball experiments, this first involved the difficult task of selecting the most suitable operative. But time-study engineeers trained by Bedaux no longer needed to start with an ideal worker, because they could effectively rate the speed at which any operative was working. They compensated not only for the differences between individual workers, or for the differing effort made naturally by the same worker at various times of the day, but even for the deliberate "soldiering" to which Taylor so often refers. In the Bedaux sixty/eighty rating, the normal rate of working (comparable to the effort of walking at three miles per hour on level ground) is sixty Bs per hour. (The "B" is derived from Bedaux.) But the expectation is that a worker of average skill and experience could work at the rate of eighty Bs per hour, and the related system of incentive payment is based on his doing this.

The expectation was not, of course, that the operative would work continuously at eighty Bs per hour, and a relaxation allowance always compensated for the conditions of work. Here again, Bedaux improved considerably on the two rest periods of ten minutes each session that Taylor allowed to the girls inspecting bicycle balls. H. E. Kearsey's historic paper on "The Bedaux Work Unit Method," first published in 1934, reports that "this relaxation allowance is not a standard allowance, but varies with conditions of work, such as atmospheric temperature, working position, periodicity of cycle, and so forth."[7]

Bedaux—who had been operating in the United States since 1911 and became a naturalized American in 1917—formed a British company in 1926 for the purpose of training time-and-motion-study engineers for his clients. One of his clients was Huntley and Palmers, where Kearsey was responsible for the introduction of work study and where it seems likely that through his efforts the company was the first to adopt the term *work study*, in 1943.[8] Other clients of Bedaux included J. Lyons and Company and Imperial Chemical Industries. Imperial Chemicals later developed one of the largest work-study units in England. Some of its officers played a leading role in founding and developing the organization that was to be called the Institute of Practitioners in Work Study, Organization and Methods—now the Institute of Management Services—to which the duke of Edinburgh devoted a five-year term as president. The American parallel to this organization is the American Institute of Industrial Engineers.

After Bedaux's two main refinements on Taylor's work, perhaps the next most important development in time study between the wars was the new sampling technique introduced by L. H. C. Tippett, used to reduce the time employed in making observations. In this case, the studies were made in England rather than in the United States. Tippett worked in the English cotton mills; the method that he used—although then called "snap reading method"—was a statistical procedure comparable to that employed in Gallup polls. It is now known as "activity sampling" and can be particularly useful when a single time-study engineer is routed to take a cycle of observations in a predetermined order. After carrying out a series of such cycles, he will be able to calculate with remarkable accuracy during what percentage of time men or machines are, or are not, employed in specified operations. Activity sampling—like the rating and relaxation allowances of Bedaux—is still with us.

THE GILBRETHS: MOTION AND METHOD STUDIES

Running concurrently with Taylor's time studies, which sought to measure a "fair day's work," are the Gilbreths' motion and method studies—motion study is a component of method study—that sought to discover "the best way" by which a task could be carried out.

Frank Bunker Gilbreth (1868–1924), who had been trained as a bricklayer, and his wife, Lillian M. Gilbreth, who trained as a psychologist, worked as a unique team. Lillian successfully continued the work long after her husband's death. They were the great pioneers of method study. In his youth Gilbreth had effectively applied his ideas concerning motion economy to bricklaying; these were later published under the title *Bricklaying System* (1909). Under the heading of "Motion Study" he says: "The motion study in this book is but the beginning of an era . . . that will eventually affect all of our methods of teaching trades. It will cut down production costs and increase the efficiency and wages of the workman . . . . To be pre-eminently successful . . . a mechanic . . . must use the fewest possible motions to accomplish the desired result."[9] Gilbreth may sound presumptuous in so discussing a trade whose standards had evolved over several thousand years, but he more than made good his claim.

Gilbreth's studies included numbering and describing the individual motions that by long tradition were involved in the single job of laying a brick. This is the third milestone—in a series of "subdivisions" of labor—contributing to the rationalization of labor and technology. The first—which appears as early as book 10 of Plato's *Republic*—was the division of labor into trades; the second was the subdivision of particular trades (as in pinmaking and clockmaking) documented in the work of Perronet and in the *Encyclopédie* during the eighteenth century; and the third is Gilbreth's analytical documentation of a bricklayer's motions while performing one clearly defined element in his work.

As in the *Encyclopédie*, each element—here motion—is numbered and described. Gilbreth provides four key charts dealing with the four methods of bricklaying that had all traditionally used eighteen motions for laying each brick. These are listed under the heading "The Wrong Way." Under the heading "The Right Way" are listed those motions that reduce the four methods for laying a brick from eighteen to four and a half, two, four and a half, and one and three-quarters motions respectively (*Writings*, pp. 56–65). As Gilbreth points out in his later publication *Fatigue Study* (1916), the elimination of unnecessary motions in bricklaying "enabled this same bricklayer to lay three hundred and fifty bricks per hour, where he had laid one hundred and twenty bricks per hour before" (*Writings*, p. 306).

Taylor frequently refers to Gilbreth's work on bricklaying. He points out that Gilbreth's success lay in eliminating "certain movements which bricklayers in the past believed were necessary"; in introducing "simple apparatus, such as his adjustable scaffold and his packets for holding bricks, by means of which, with a very small amount of cooperation from a cheap laborer, he entirely eliminates a lot of . . . time-consuming motions"; and in teaching "his bricklayers to make simple motions with both hands at the same time" (*Principles*, pp. 77–85; *Testimony*, pp. 66–73).

THE RECENT GROWTH OF SYSTEMS AND SOCIETIES

The Gilbreths were remarkably adaptable in the use of interdisciplinary methods for furthering their studies in motion economy. Gilbreth later analyzed all jobs into seventeen basic motions, which he called Therbligs, an anagram of his name. In analyzing and documenting the motions, he used motion-picture photography, then in its infancy, and employed cyclegraphs and chronocyclegraphs to measure activity. Making use of his wife's special training, Gilbreth was able to employ both physiology and psychology in studies that made him one of the early exponents of ergonomics, which improves the effectiveness of a worker through the design of the work setting.[10] The Gilbreths also influenced other developments in motion study during the second and third quarter of the twentieth century. The new developments included Predetermined Motion Times (PMT) and Methods-Time Measurement (MTM), which provide synthetic times for a whole series of activities, thereby permitting the time required for a new job to be predicted.[11] These, in their turn, contributed to a more recent development, the British Work-Measurement Data Foundation.[12]

One of the most confusing aspects of work study results from the way in which related systems, and hence their titles and terminology, have mushroomed since World War II, frequently using methods developed by the armed services. Operational Research and Games Techniques speak for themselves; ergonomics was furthered by the need for aligning men and women with sophisticated machines; CPM (Critical Path Method) "grew out of a joint effort in 1957 . . . to apply electronic computers to scheduling the design and construction of chemical plants"; PERT (Program Evaluation and Review Technique) was designed in 1958 by the United States Navy's Special Projects Office "for evaluating the . . . existing schedules on the Polaris missile programme";[13] and cybernetics is "the study of *control systems* in man and machine," which may be compared with the feedback supplied by the Watt governor in its control of steam engines since the early part of the industrial revolution.[14]

Time and method were not altogether ignored in clerical and administrative work before the twentieth century. In the eighteenth century, James Watt, for example, invented not only the steam engine but also the copying machine in order to duplicate his own business letters and engineering drawings. In the nineteenth century, Anthony Trollope, the novelist—like Charles Dickens, "in all things as punctual as the clock at the Horse Guards"—demonstrated that time measurement can also apply to creative work. As Trollope said, it had "become my custom . . . to write with my watch before me, and to require from myself 250 words every quarter of an hour. I have found that the 250 words have been forthcoming as regularly as my watch went. . . . I wrote my allotted number of pages every day. . . . And as a page is an ambiguous term, my page has been made to contain 250 words."[15]

The idea that creative writing can be subject to time study has not been generally accepted, even in the twentieth century. But office work—as typified by Watt's copying machine, Babbage's calculating engine, and the typewriter of Sholes and Remington—has become more and more a subject of method study within our own lifetimes. When I was a young man, office workers still argued heatedly that method study could only be applied to industrial work, hence the American term "industrial engineering." In recent years that myth has been exploded with the aid of computers, word processors, and MBAs. As Currie puts it, in his *Work Study* (1963): "The old concept of method study as applying only to light repetitive work did scant justice to its potentialities. It can, in fact, be applied anywhere, since any process or procedure is open to improvement. Fundamentally, method study involves the breakdown of an operation (or procedure) into its component elements and their subsequent systematic analysis. Thence, those elements which cannot withstand the tests of interrogation are eliminated or improved."[16]

During our century there has been an increasing interest in the rationalization of operations and procedures. Concurrent with the growth of the two major time-and-method-study institutes in Britain and North America—the Institute of Management Services and the American Institute of Industrial Engineers—has come a virtual deluge of publications on every aspect of productivity science. Work study, like the philosophy of Machiavelli or Hobbes, is as frequently berated as it is employed. There is no need to document the reaction to the rationalization of production. It comes as naturally to the trade unions today as it did to the weavers in Thomas Shadwell's *Virtuoso* (1676) and Gerhart Hauptmann's *Die Weber* (1844). But work study is no more than a logical system of procedures for enhancing efficiency in all areas of human endeavor. It can hardly be blamed for the use to which some members of society—for the most part employers—have put it in the past.

Perhaps the greatest single problem facing men and women in the last quarter of the twentieth century is that of apportioning—both among nations and between humans—those material goods that the technological revolution has brought within the reach of all.[17] In his conclusion to *Principles of Scientific Management*, Taylor not only clearly foreshadowed the problem but recognized that work-study practitioners (by whatever name, including the recent proliferation of "management consultants") would play a leading role in apportioning the profits from scientific management between the workers, the employers, and the consuming public. He states that "the third great party, the whole people—the consumers who buy the product of the first two and who ultimately pay both the wages of the workmen and the profits of the employers"—must insist on a proper division of those profits.

In a world of endemic inflation, plagued by lockouts and strikes—where workers are often paid according to their ability to disrupt rather than their willingness to produce, and where labor and management are continually coming to terms at the expense of the consuming public—Taylor's words have an ominously modern note. Though he deplores employers who attempt to drive workmen "into harder work for lower pay," and labor "which demands one increase after another... while at the same time it becomes less instead of more efficient," Taylor is ultimately optimistic:

> The means which the writer firmly believes will be adopted to bring about, first, efficiency in both employer and employé and then an equitable division of the profits of their joint efforts will be scientific management, which has for its sole aim the attainment of justice for all three parties through impartial scientific investigation of all

the elements of the problem. For a time both sides will rebel against this advance. . . . but in the end the people through enlightened public opinion will force the new order of things upon both employer and employé. (*Principles*, pp. 136–39)

Taylor's words are remarkably prescient concerning the perceived central malaise of our times, the problem of sharing wealth not only within but also between nations. Will we ever make the effort to ensure that his optimism is justified?

THE RELATIONSHIP BETWEEN WORK STUDY AND MANAGEMENT

Time and motion study first developed as part of the function of the owner-manager who was intent upon improving the efficiency of his organization. The growth of large corporations and government-sponsored undertakings was part of a process through which the ownership of capital has become to a great extent divorced from management. In our own century management has itself broken down into more and more areas of specialization, in order to improve its efficiency. This specialization can be regarded as a division of mental labor. The process is analogous to the division of physical labor that so dramatically improved the efficiency of manual production from the second half of the seventeenth century. The division of mental labor has produced a plethora of management-related functions grouped under such headings as production manager, office manager, works manager, transport manager, stock manager, accountant, sales manager, and purchasing officer, not to mention hospital management, educational administration, and so forth. The management-related functions are, in their own turn, creating the need for numerous professional and university qualifications.

Work study, too, is only one of many functions of management that have been recognized as discrete entities as a result of the division of mental labor. But work study is unique in providing a tool that can help all managers to manage the workers, the tools, and the plant. This is not all. Work study will inevitably become involved in an increasingly critical examination of the organization and methods in management itself. After all, inefficiency in a manager will generally be far more damaging than in a secretary.

Just as office workers, until quite recently, mistakenly believed that their routines could not be subjected to the techniques of work study, so some managers still consider that their functions are sacrosanct. More than one work-study engineer has been faced with the problem of persuading a manager to organize his time, delegate his power, and list his priorities. If thirty years of association with work study have taught me anything, it is that none of us are immune to the analytical questioning with which these techniques are ultimately involved. Work study does not stop at the door of the manager's office, the desk of the author, the table of the boardroom, or even the procedures of the work-study engineer. Work study—in all its forms and by all its names—is an integral part of the restless questioning and seeking after material progress that, since the latter part of the seventeenth century, has produced an exponential growth in Western technology. And this is the technology in which, for better or for worse, the whole world would now like to share.

THE INTERNATIONALIZATION OF RATIONALIZED PRODUCTION METHODS

The industrialist who best illustrated the use of Taylor's Scientific Management was Henry Ford (1863–1947). As David Halberstam puts it, "Ford, fascinated by efficiency of production, absorbed Taylor's principles and began to use them in his plant, eventually developing and applying them to an almost mythic degree."[18] Ford is best known for first introducing the assembly line in 1913 to manufacture automobiles. But he also introduced into this procedure the standardization of products, the integration of supply industries, the dispersal of assembly plants, and the organization of manufacturing around a continuous line-to-line flow of product components.

Starting with virtually no capital in 1903, Ford gradually bought out the other stockholders. In 1927—when he had built the last of more than fifteen million Model Ts since 1908—the Ford Motor Company had a surplus of nearly seven hundred million dollars. Ford's industrial philosophy went one step beyond the normal pace of Western technology. Instead of waiting for demand to increase through rationalized production methods, he first reduced the price. This increased the volume of sales, thereby permitting him to improve the efficiency of production, increase volume further, reduce price further, and continue this cycle almost indefinitely.

Despite his paternalism, Ford was genuinely interested in the well-being of his employees. He was also a man who could build a car with his own hands. In 1914—when the average wage in U.S. manufacturing industries was eleven dollars a week—he paid five dollars a day, though he expected value for money. Ford hated banks, Wall Street, and trade unions. He financed his company as far as possible out of profits, and hoped that unionism and its proponents would pass away. But in 1941, when he lost a union-recognition election, which was conducted by the National Labor Relations Board, the United Auto Workers received from him the first union-shop and union-dues-checkoff contract in the automobile industry.

Like Matthew Boulton—of Boulton and Watt, who built the first great factory for steam engines in England over a century earlier—Ford was a natural inventor of mechanical products. As one might expect, Boulton and Ford both had frustrated dreams about the mass production of clocks and watches. Boulton had great hopes for such production at his Soho Manufactory in the period 1769–75, and Ford had plans to produce six hundred thousand watches a year at thirty cents each.[19]

Despite his reputation as a trail blazer, Ford had some very rigid views. They were a source of strength in the early days but eventually cost him dearly. He tried to retain the mechanical brake, the planetary transmission, and the four-cylinder engine, rather than the hydraulic brake, the conventional gearshift, and the six- or eight-cylinder engine. He also continued to insist that the Model T could be purchased in any color, as long as it was black. Ford at last realized that he had been too tardy in making changes. He retooled completely for the Model A by 1928 and produced a V–8 by 1932. Yet it was too late to regain his dominating lead. In 1920 Ford had been manufacturing half of all the motor vehicles in the world; by 1928, though it remained firmly in second place, Ford had fallen behind General Motors.

General Motors had also started in 1908, when Ford made its first Model T. William Crapo Durant, who had used profits from a successful carriage-making business to found General Motors, was quite a different man from Ford. Much of his attention was devoted to the stock market. He lost control to a banking syndicate in 1910 and then took over again in 1916 with profits that he had made out of building up Chevrolet. Durant was finally forced out after the financial panic of 1920. By 1923 General Motors was firmly in the hands of Alfred P. Sloan, Jr., under whom it became the largest manufacturing enterprise in the world. The difference between the managements of Ford and General Motors was in many ways the difference between the old-fashioned business proprietor of the eighteenth and nineteenth centuries and the large public company with one eye on Wall Street (and the bottom line) that we know today.

Aldous Huxley wrote his dystopia, *Brave New World*, in 1931. Since he wanted to undercut the completely rationalized society into which he felt that we were developing, it was natural that he should choose "Our Ford" as the modern equivalent of "Our Lord." It was equally natural that the inhabitants of such a world would make their obeisance with the sign of the T. One of the further allusions to rationalization, which seems to have escaped critical notice until a recent article by James Sexton, appears in the name Mustapha Mond (must-have-a-world) that Huxley gave to the World-Controller. According to Sexton, Mustapha Mond corresponds in both appearance and nature to Alfred Moritz Mond. In 1926 Alfred Mond had been a founder of the huge Imperial Chemical Industries, which is still a British leader in the use of work-study methods. He had been made Baron Melchett in 1928 and died in 1930. Lord Mond was a leading exponent of industrial rationalization who provided the following definition of that term for the 1929 edition of *Nuttall's Standard Dictionary*: "The application of scientific organization to industry by the unification of the processes of production and distribution with the object of approximating supply to demand."[20]

Yevgeny Zamyatin, a naval architect and mathematician, was the dystopian writer most directly concerned with the influence of the clock on Western technology. He makes Taylor, and "the Taylor system," the most important single target in his novel *We* (1920).[21] By the mid-twenties Ford, the greatest exponent of the Taylor system, had produced at River Rouge the most perfectly rationalized and integrated factory complex in the world. Its eleven hundred acres accommodated some seventy-five thousand workers and twenty-seven miles of conveyor belts. Halberstam reports that a British historian said of the

Rouge: "Here is the conversion of raw material to cash in approximately thirty-three hours." Sixty years later, Toyota would be credited for the JIT, or just-in-time theory, which reduces inventory by relating the supply of parts directly to manufacturing. But this aspect of industrial rationalization, like much else, had already been applied at the Rouge. As Eiji Toyoda, of the Toyota company, said, in toasting the head of Ford in 1982, "There is no secret to how we learned to do what we do, Mr. Caldwell. We learned it at the Rouge."[22]

In the spirit of the story of the fox and the sour grapes, the North Americans and the British have tended to stress that the Japanese are better at imitation than invention. But both countries ignore their own history. We have seen how, in the latter half of the seventeenth century, the British took over clockmaking and indeed much else from the Dutch. In the nineteenth century the Americans used their waves of cheap immigrant labor to do much the same with the British. Dryden pointed out that the genius of the British is "rather to improve an invention than to invent themselves." Defoe says much the same in his *Plan of the English Commerce* (1728): "It is a Kind of Proverb attending the character of *English* Men, that they are *better to improve than to invent.*"[23] If history can teach us anything, it is that the Japanese, like the British and the Americans before them, will start inventing once they run short of products to imitate.

The real strength of the Japanese lies in the fact that, whereas Western technology has needed three hundred years of free enterprise to evolve, it takes barely two generations to transfer. Also, free enterprise has spawned unacceptably high expectations of material rewards from both the management and trade unions of Britain and North America. As predicted by Taylor, management and unions frequently enter into an implicit compact at the expense of the public. But the Japanese have no such traditions. Their implicit compact is feudal. The feudal relationship is still best defined in the *Germania* of Tacitus (ca. 55–117 A.D.). As he explains of the Germanic tribes, this is the compact of what the Germans know as "Milde und Treue": the leader owes generosity to his followers and the followers owe loyalty to the leader. Such relationships no longer play any part in our Western industrial structures.

But armaments and war still play a vital role in Western industry. During World War I, America turned from a debtor into a creditor nation at the expense of the British. In World War II, America became by far the greatest exponent of rationalized production that the world has ever witnessed. The United States built nearly three hundred thousand military aircraft during that period. Henry Kaiser, who had dreams of taking over the mantle of Ford, was reputed to be launching one of his ten-thousand-ton Liberty ships every day. Even in the period after the war it seemed as though rationalization would keep the pot of gold continually full. By 1955 General Motors had produced its fifty-millionth car, and in that period the auto industry accounted for almost one-fifth of America's gross national product.

But the tide of prosperity would turn. Motivated in part by an altruism unprecedented in the history of civilization and in part by a morbid fear of communism, the United States underwrote the rebuilding of West Germany and Japan. They rose like twin phoenixes from their ashes. Like London after the cataclysms of the Great Plague and the Great Fire, West Germany and Japan produced a better base for the development of production and technology precisely because they were making a clean start. Both worked particularly hard to build up their steel and automobile industries. Like Britain, their traditions lay with the small car. But, unlike Britain, they did not suffer from an implacable enmity between workers and management—"us and them"—which was constantly reflected in poor deliveries and unreliable quality. The Americans certainly had warning enough about the need to improve quality, to avoid constant changes in design, and to retool for the production of smaller cars. Yet, in the period leading up to 1974, the American automobile industry ignored the warning signs, just as Ford had done in the period leading up to 1927. Apparent success can be a dangerous sedative.

In the postwar years a mainstay of American policy had been to hold down the price of gold to thirty-five dollars per ounce and the price of gas to about thirty-seven cents per gallon. In the early seventies—as a result of difficulties with the balance of payments—the United States lost control of the price of gold, thereby giving joy to both South Africa and Russia. Gold has since then jumped more than twenty-fold in price, before settling back somewhat. Starting in 1974, OPEC was able to take advantage of the Yom Kippur War of October 1973 to raise the wholesale price of oil by a factor almost as great as that of gold. The results have been disastrous. The Western world began to hemorrhage almost immediately—an image applied by Defoe to bourgeois types who spend more than they can earn.[24] Even Americans began to realize—at least for a time—that large private cars should long ago have followed the dinosaurs. But Detroit was too late to retool effectively. Its management was fat and its workers were recalcitrant and overprotected. By 1975 the small imported passenger cars had taken almost one-fifth of the American market, and the public loved their reliability and quality.

As Halberstam notes, the coming of Ford had coincided with the discovery of vast reserves of oil in the American Southwest. This meant that the United States became the one industrialized nation with cheap oil.[25] But after 1974 the Americans could no longer continue to complain that Japan was benefiting from either cheap Arab oil or cheap indigenous labor. With a more motivated and disciplined labor force, Japan was able to overcome the fact that, unlike America, it had to import all of its oil with foreign currency. Moreover, the marked rise in the value of the yen during the latter part of the eighties made the Japanese worker the highest paid in the world.

From the point of view of making rationalization work, Ford was probably right to pay his workers well. Whatever technology one uses, it takes only a few employees to obstruct deliveries and diminish quality. In the past, sixty-nine grain handlers in Prince Rupert have, during the winter months, been able to hold up 30 percent of Canadian grain exports. Large Japanese corporations, by promising "lifetime" employment to about a third of their employees (generally males until the age of fifty-five), have ensured a stable cohort of loyal and adaptable workers. Such workers have no interest in obstructing the new technology or the considerable flexibility that derives from contracting out to small factories both at home and abroad. Though the Japanese have recognized unions since World War II, they are one-company unions that appreciate the primacy of their firm's interests.

For a time in mid-1985 it looked as though OPEC might break, and by the spring of 1986 oil had fallen to about ten dollars per barrel. The price of oil has, however, rebounded since then, and the Western nations, which continue to use large quantities of imported oil, appear to have learned very little. There seems also to be a distinct difference between the way in which the Arabs and the Japanese have employed the many hundreds of billions of dollars that they have extracted from the American economy. Some understandable reticence exists with regard to examining directly what has happened to the very large sums that have flowed back into the West from the Arab world. Clearly, however, comparable sums were then being loaned out by Western banks—particularly in the United States, Canada, and England—to Third World countries whose creditworthiness seems hardly to have been investigated. In Canada, for example, these loans by the major banks greatly exceeded their total capital, yet neither the bankers nor the politicians have been investigated. The term "sovereign loans" is now being applied euphemistically by bankers, whose greed is costing the Western nations dearly. The bankers are far from having learned from such mistakes, and their current LBOs or leveraged buyouts—using our pension and other funds for corporate takeovers—may well prove equally disastrous for all concerned.

The Japanese, who have worked rather harder for their money than the Arabs, are now becoming bankers to the world. Thus far, they have generally been able to restrict their activities to those countries in which they consider it prudent to invest. According to Peter Spry-Leverton and Peter Kornicki, "In 1984 the Japanese share of the international banking market was 23 per cent, while the United States share was 26 per cent and the British 8 per cent. By September 1986 . . . the Japanese share had risen to 31 per cent while the American and British shares had fallen to 18 and 6 per cent respectively." By 1988 all of the ten largest banks in the world were Japanese. Like the Dutch in the latter part of the seventeenth century, when faced by the rising industrial power of the British, the Japanese seem to be beginning to consolidate their position. They may perhaps now rely as much on finance as on the direct advantages of trade. One might even suggest that their recent move into such advanced research as fifth-generation computers indicates that some of their present industrial advantage is being lost. With the rise in the price of the yen, Japan is already facing very stiff competition from the so-called four little dragons: South Korea, Taiwan, Singapore, and Hong Kong (by 1990 their joint exports are expected to exceed Japan's).[26] Like Japan before it, South Korea has carefully organized an assault on world markets. Their program has been based on a productive work force and was spearheaded by steel, shipping, and car manufacturers. And like Japan, South Korea is now moving into a whole range of electronics, with which it proposes yet again to use the advantage of its cheap and productive labor.

There is a curious parallel between the changing centers of prosperity in Britain and the United States, which has much to do with the productivity of labor. In seventeenth-century England the power of the trade guilds was concentrated in the City of London. After the Great Fire tradesmen began to move outside the City limits—the clockmakers, for example, to Clerkenwell—in order, in part, to evade the control of the guilds. In the eighteenth century large manufactories were increasingly set up in the North. There, water power, coal, and cheap labor were increasingly available, with even less control by the guilds. In 1700, leaving aside London, there were only two towns—Bristol and Norwich, both also in the South—with some thirty thousand inhabitants. By 1800 there were fourteen such towns, and virtually all of the development had occurred in the North. Towns like Manchester, Salford, and Birmingham had grown from villages into urban centers with up to eighty thousand people. For the most part, women and young children, as well as men, worked under the most abominable conditions, after leaving the type of countryside that is nostalgically portrayed in Goldsmith's *Deserted Village* (1770). These conditions created the British antagonism between "us and them." Today, prosperity has, once again, moved to the South. Few industrialists seem eager to set up new industries in the North, where labor problems appear to be endemic.

Something similar has happened in the United States. In the early years, slave labor provided plantation owners in the South with a very comfortable life. As Defoe's novels demonstrate, Englishmen could expect to make a remarkably good living there from very small beginnings. But after the Civil War the cheap labor previously provided by southern blacks began to be less effective, even for wages. In the North, however, wave after wave of cheap European labor, as well as former slaves from the South, provided the underpinnings for America's great period of industrial hegemony. This reached its peak in the period around the two Great Wars. But since that time America's labor and management, centered in the North, have become too inflexible and too much featherbedded. As a result, industrial action is now beginning to be concentrated in the South. There—as in the North of England during the eighteenth and nineteenth centuries—the power of trade unions is far less in evidence. The United States has also seemingly been involved in a "closed eyes" attitude toward the influx of several million illegal Hispanic immigrants.

It would seem at present that, if industry is to become reinvigorated in the United States, the South may be the main area of growth. Indeed, foreign companies, like those from Japan, are already setting up there. But international corporations, and Americans remain strongly entrenched in them, are not the exclusive domain of any single country. In a more recent development, *maquiladoras*, or twin-plant facilities on the Mexican side of the border with the United States, supply labor-intensive components at labor costs that, in 1988, were said to be about $2.30 U.S. a day, compared with about $15 U.S. a day in Hong Kong or Taiwan. It is estimated that in 1986 American companies were employing some 250,000 workers in Mexico alone. Japan has resisted the potential invasion of what the Americans euphemistically call "joint ventures," but South Korea has been more accommodating. Such American ventures are now to be found not merely in the West but in all areas of the Third World where governments are cooperative and labor is cheap, disciplined, and adaptable. We may take it for granted that, though the Americans are still the leading exponents of offshore manufacturing, the Japanese will not be far behind.

Early in the period with which we are concerned, Western manufacturers—as we shall see in chapter 10—began the effective exploitation of cheap indigenous labor. The history of Britain since the killing of the king in 1649 has been the history of a transfer of power to the common man and woman. But liberalism and the spread of the franchise has increased the cost of indigenous labor and made Western manufacturers look toward the cheaper workforce in less-developed countries. The older dichotomy between the "haves" and "have-nots" within Western nations is now changing to a dichotomy between Northern "haves" and Southern "have-nots" in our global village as a whole. Yet the spreading of technology to Southern lands will inevitably bring with it the same distribution of the franchise that has thus far been restricted to Western countries.

Ultimately, there is no way to hive off politics from the proliferation of technology and the related distribution of franchise and wealth. But the transition period is generally painful. Our very terms *right* and *left* in politics derive from the seating of political representatives in the French government. Under the word *center*, the *Oxford English Dictionary* explains that in the French Chamber, arranged as an amphitheater, those of moderate opinions occupied the central benches: "This use originated in the French National Assembly of 1789, in which the nobles as a body took the position of honour on the President's right, and the Third Estate sat on his left. The significance of these positions, which was at first merely ceremonial, soon became political." Those who wished to continue the change away from aristocratic privilege remained on the left as radicals, while those who wanted to hold back on change moved to the right as reactionaries.

The British and American traditions have not been associated with the same physical dichotomy as the French, but a parallel sense has developed of a right, which is reactionary or conservative, and a left, which is radical or liberal. Frequently the political persuasions of people, of classes, and of nations reflect where the speakers or their representatives stand on the spectrum beween haves and have-nots. It has been said that a man of twenty who is not a socialist has no heart, and a man of forty who is not a conservative has no head. But it is also true that young men can afford to preach the need to be liberal, while older men have an interest in conserving what they have. Frequently, too, those who are loudest in preaching the division of wealth are people or nations who feel that they will gain from a more equal distribution.

Yet, in the world of increasing equality described in this book, there are already those who are beginning to sense that a drawing together of the left and the right in the direction of the center has merits of its own. The traditional roles of Republicans and Democrats—which reflect respectively the right and the left in American politics—are already beginning to blur; Canada even has its Progressive Conservatives. But most Westerners have difficulty in recognizing the extent to which the Communists of both China and Russia have recently been trying to emulate the West. Not merely Western clothes and tastes but even Western political ideas are being imitated by people and governments who hope that they, too, might share in the Western cornucopia of material goods. At the same time the United States is becoming heavily involved with government support systems—such as social services and farm support—that have hitherto been associated with the socialists.

Despite adopting their Bill of Rights as long ago as 1791, Americans have been very slow to demonstrate that blacks, Jews, women, and a whole spectrum of non-Anglo-Saxon immigrants were really created equal. But eventually a true franchise has developed because free enterprise has demanded an increasing equality of opportunity together with an outlet for its inundation of consumer goods. The Communists had also declared equality for all comrades at the outset, but their real record on distributing both the franchise and consumer goods has been much worse than that of the West. At last, however, they are beginning to realize that they can no longer keep the seductive Western lifestyle at bay with political dogma.

The historic Communist party conference in Moscow in July 1988 reflected these changes and the fact that equality of opportunity in the political and economic areas are interdependent. The predominant clothing at the conference was standard Western in style. For the first time there was open debate on television. It displayed all the cut and thrust of Western conservatives—who are reluctant to reduce entrenched privileges—defending themselves against Western liberals. Both the Americans and the Russians had insisted that there was no place for party politcs in Russia. But television viewers all over the world witnessed an embryonic party of the right and of reaction being led, however tenuously, by their champion Yegor Ligachev. They were defending themselves from the forces of the left and of change epitomized by Boris Yeltsin. Resignations of several of the old guard, including Andrei Gromyko, were even called for by name.

In the middle, trying to hold the center ground, was a suave Mikhail Gorbachev. He even tried to obtain a limitation of his own powers by calling for no more than two five-year terms for any president, a procedure that patently imitates the American tradition. He was also attempting to introduce—at all political levels, and for the first time—the real franchise (meaning the real equality) of one *person*, one vote. Gorbachev recognized the inevitable corollary of a separation of "church" and state, which in his country meant the separation of the Communist party and its dogma from the political process. Naturally, forces of privilege—like the eighteen million bureaucrats and the members of the Communist party—will engender strong pockets of resistance. But, sooner or later, the spreading fruits of Western technology will make real equality of opportunity and a real franchise as inevitable in the Communist world as it has been in the West.

In Western Europe the impending standardization is distinctly more comprehensive than the free-trade arrangements between the United States and Canada. Britain, the pioneer of rationalization, is now becoming rationalized itself as part of the European Community. By

1992 the EC is scheduled to remove thousands of internal trade barriers, to free capital flows, to make excise and value-added taxes more uniform, to open all bidding for contracts within the EC, and to ensure that workers can practice their trades and professions anywhere in the Common Market. Of course, there will be reaction from those with vested interests, but a United States of Europe with 325 million inhabitants no longer seems impossible. Furthermore, the twelve nations of the EC have—in Portugal, Spain, Italy, and Greece—a built-in equivalent of North America's cheap Hispanic labor force.

The process of industrial rationalization is assuming world proportions before our very eyes. Commensurate with this change is an increase in education and a slow but sure entry of more people into the ever-growing middle class. As a result there is a continually increasing demand for just those material goods that the international corporations produce. Between 1960 and 1982 the proportion of the population in India aged twenty to twenty-four enrolled in higher education rose from 3 to 9 percent. India has now reached the same percentage as Britain had in 1960. Moreover, the pool of scientists and engineers in India had grown from 190,000 in 1960 to 2.4 million by 1984. This increase in the education in what the World Bank calls lower-middle income areas is a powerful indicator of where our world may be heading. It suggests that, in addition to some 1.1 billion people who already enjoy the material advantages of Western technology, perhaps an additional 2.6 billion people may be ready for the transfer of technology within two to four generations.[27]

There is probably an even better and simpler test for measuring what is happening in the world than the rise in literacy and education. The increased abidance, worldwide, to accurate time measure provides another indication of the new rationalization. Thomas Tompion—the father of English watchmaking, who died in 1713 after introducing batch production—produced the impressive quantity of 6,000 watches in his lifetime. At the turn of this century, Ford discarded his plans for the mass production of cheap watches, because he could not envisage the necessary sale of 600,000 pieces per year. According to the Citizen Watch Company of Japan, the total worldwide sale of watches in 1986 was just under 530 million pieces, of which the Japanese provided 190 million. The sale of quartz watches alone had risen to 395 million in 1986 from 241 million in 1983; mechanical watches had dropped from

140 million to 135 million during the same period. The worldwide sale of watches seems now to be rising at the rate of 14 percent per annum.[28] Technology is a far more efficient equalizer than the world has thus far recognized. Nor have we fully understood what these figures tell us about the impending rationalization of both our 5 billion people and the material goods that they will increasingly demand.

IX Retail Distribution and Finance

We should now look briefly at the role of rationalization in distributing goods by retail and in financing production. At the beginning of the period in England with which we have been chiefly concerned, farm products were generally brought to urban markets by their growers, and foreign products were imported by spice, wine, and other merchants. The products of the guilds, however, were normally sold in the shop at the front of the house in which the master lived and worked. Thus, for guilds and for farmers, producing and retailing were generally two sides of the same coin.

The agricultural revolution in England used the enclosure of the common lands to rationalize both the methods and the structure of farming. When the movement began, perhaps as many as 95 percent of people lived and worked in the country. Today the numbers have been completely reversed, and the same percentage or more live or work in urban surroundings. When we consider the current displacement of workers through technological change, we should remember that most of the displaced agricultural workers did find work in the towns, however difficult their lot. In the latter part of the eighteenth century, the agricultural revolution sent wave after wave of farm laborers into the towns. There they provided cheap labor for the burgeoning factories, much as the waves of cheap labor from Europe would fuel the factories of America a century later. The guild system did not break up immediately, but it was considerably weakened by the production of goods outside its control.

RATIONALIZING RETAIL DISTRIBUTION

As the production of goods came to be increasingly divorced from retail distribution, there developed a greater degree of specialization in the retailing of particular products. Such specialization included men's outfitters, drapiers, or hardware merchants. But in some ways the pattern followed that of the guilds. A young man would learn a particular trade, and then—much as in the guild-type apprenticeship system of Hogarth's *Industry and Idleness*—if he had applied himself well he might one day become a shopkeeper himself. By the turn of the nineteenth century, Napoleon could, with excellent justification, refer to England as "a nation of shopkeepers."

Similar trends, though on a much larger geographic scale, were taking place in the United States. In the nineteenth century more and more commercial travelers and agents, representing both manufacturers and wholesalers, were in regular contact with retail stores. Perhaps because the distances to be covered were so great, the American "traveling man" became part of the national psyche. We see this not only in Arthur Miller's *Death of a Salesman*, but also in the way that the real traveling-salesman fathers of Eugene O'Neill and Tennessee Williams are reflected in their plays. During the nineteenth century the number of retail stores multiplied like the population, in towns all over America. By 1914 chain stores specializing in particular lines were beginning to replace the general store. At the first census of retail establishments, in 1929, there were found to be 1,543,138 stores with total sales of just over 49 billion dollars. Though the sales were cut almost exactly in half during the depression, the figures had bounced back to 1,770,355 stores by 1939, with total sales of 42 billion dollars.

By 1948, however, a smaller number of American establishments with sales of just under 129 billion dollars confirmed a most important trend in retail distribution. The trend had begun in the latter part of the nineteenth century. In 1879 Frank Winfield Woolworth set up the first of his "five-and-ten-cent stores" in Lancaster, Pennsylvania. In this store he offered a larger selection than he had in an earlier, failed attempt. A number of Woolworth's relatives and close friends started similar stores of their own. These were operated independently until 1912, when they were merged into the F. W. Woolworth Company.

When Woolworth died, in 1919, more than one thousand stores were being operated by his company.

By 1910 Woolworth had spread to Britain, but in that country there were already indigenous developments of a comparable nature. By 1900 one could find multiple shops of grocers, provision merchants, tea dealers, newsagents, tobacconists, and chemists. In this century some of the better-known names have been W. H. Smith, Lyons' tea-shops, and Boots, the chemists. For retailing a wider range of goods, there were also penny bazaars, and Marks and Spencer, which was founded in 1887. The retail cooperative societies in Britain had grown to the remarkable membership of 12.5 million by 1960, but they have lacked the competitive spirit necessary to hold their own during recent years. What the free-enterprise chain-store operations tended to have in common was a centralized office, which generally included a varying amount of centralized buying. But retail chain stores were also rationalized in another way. They tended to be developed as clones, patterned after a successful prototype whose management, staff, stock, layout, and modes of operation were duplicated.

In many ways the corporate chain store was the first successful method for the large-scale integration of a system for retail distribution. In the Britain of 1950 they handled more than 23 percent of the total annual retail sales of just under six billion pounds. As in the United States, they continued to grow throughout the sixties. Their use of brand names (their own and those of others) has further influenced the rationalization of retailing, both in food products and elsewhere. They also maintain quality standards that are recognized across the country and even across the world. Marks and Spencer has developed a unique form of centralized cost and quality control over its manufacturers, which has enabled it to corner a very large part of the total sales of underwear throughout Great Britain. In the years between the two Great Wars, and for at least two decades thereafter, the red storefronts of Woolworth and the green storefronts of Marks and Spencer were to be found almost side by side at the center of any British town or suburb of respectable size.

Corporate chains have also had a major influence on the smaller stores. In attempting to compete, they have frequently formed voluntary cooperative chains. Often a central wholesaler provides its own as well as standard brands but allows the retailer to maintain his independence. The competition from the chains has frequently made the smaller stores rationalize as far as possible, in order to compete.

Until very recently, however, the productivity in wholesale and retail distribution in the United States has lagged a long way behind that in agriculture and manufacturing. The percentage of the population in agriculture has been falling for more than two centuries, while the productivity in that area has risen accordingly. There were about three times as many people employed in construction and in manufacturing in 1950 than there were in 1870. The comparable figures for the wholesale and retail trades show a remarkable ten-fold increase. This reflects an increase in turnover and added functions, but also a decrease in hours worked from sixty-six per week in 1870 to forty-four in 1949. Also, the increase in productivity was low. It has been estimated, for example, that while the increase was 1.1 percent annually per manhour in the wholesale and retail trades, it was 2.6 percent in mining, agriculture, and manufacturing.

There has been a myth, only recently exploded, that wholesale and retail distribution were, like office work, not really amenable to time and method study. Latterly, a development in retail distribution seems to presage a change at least as great as that made by the coming of the chain store. Large stores—like large organizations of all types, particularly in Britain, Canada, and the United States—have been suffering lately from labor difficulties arising from union activity. It has been one of the paradoxes of recent years that large corporations have been heavily featherbedded through government subsidies, particularly in Britain and Canada. Virtually all of the increases in employment have, however, come from organizations with under one hundred employees, and frequently far fewer than that. In Canada, for example, 96 percent of all new jobs between 1978 and 1984 came from businesses with fifty employees or fewer. One estimate, in 1987, was that in recent years three million jobs had been lost in the United States in large businesses, but that this had been more than made up by an increase of some six million extra jobs provided by small business. This almost certainly occurs because the productivity of very small companies benefits from the fact that they are difficult for unions to organize. The proprietor and his or her family may be the sole employees. In any case, he or she keeps a much closer control; in small businesses, lack of vigilance leads to bankruptcy, the commercial equivalent of death.

FRANCHISING: A NEW WAY TO RATIONALIZE THE DISTRIBUTION OF GOODS AND SERVICES

We shall deal with the spreading of the human franchise in North America (as it relates to equal opportunity) in the next chapter; what is relevant here is that greater opportunities have led to a considerable increase in potential business proprietors, particularly among women and the conspicuous minorities. Lately, virtually all new jobs are offered by small rather than large businesses. This has led, almost by a process of economic necessity, to the engendering of a new "invention": the remarkable burgeoning of franchises during recent years. There are other new selling formats—such as direct selling of the Amway type and TV selling—but their total impact is difficult to evaluate at this time. Franchises, on the other hand, combine the energy and vigilance of the small proprietor, the cheaper labor available in smaller businesses, and the chain-store advantages of the "clone" front and "clone" products backed by a centralized buying organization with know-how. The franchisee has the disadvantage that he or she must sometimes pay a fee, as well as a percentage of takings or profits, to the franchisor. In 1984 Meg Whittemore estimated that, whereas 65 percent of new enterprises fail within five years, only 3.3 percent of franchisee-owned outlets closed in 1982.[1] Figures vary, but the rate of closings for franchises is always very favorable, and this is reflected in the preferred treatment regarding credit that they receive from banks.

Though the formats of franchising vary greatly, they tend to break down into two main types. The first, known as product and trade-name franchising, has been with us for most of the twentieth century. This form of franchising provides retail outlets for such very large corporations as auto and truck manufacturing, gas service stations, and soft-drink bottlers. Such dealerships tend to carry the company name, though frequently combined with the name of the operator. In the late fifties there began the second type of franchise, which has become known as "business-format franchising." This is the area of franchising that has been burgeoning throughout the seventies and eighties. Such franchises have been increasing both in the number of outlets and in the scope of goods and services that are offered.

Perhaps the best test of business-format franchising is that the public is never really sure whether they are dealing with a corporate organization, which follows the chain-store pattern, or with a franchised operation. Business-format franchises normally start with one or more prototype outlets, and franchisors frequently continue to operate these themselves, even after selling rights to franchises. One advantage is that the practice keeps the franchisors directly in touch with the day-to-day operation of their businesses. This practice is so widespread that Evan E. Anderson has been able to assess the optimum percentage of stores that franchisors, in various trades, should operate themselves.[2] Clearly, the model for such franchises has been the McDonald's Corporation, based on a prototype and founded in 1955. By 1970 the chain had 1,500 outlets, and in 1985 this had grown to about 8,500 units, of which 500 had opened in the previous year.

John Naisbitt estimated that by the year 2010 the sales of business-format franchises alone should have grown from the 1985 figure of 140 billion dollars to some 1.3 trillion dollars. But by the spring of 1988 there were already 4,000 franchisors in the United States with over 500,000 franchises and .5 trillion dollars in sales. In Canada, which is also the preferred foreign outlet for American franchisors, there were then 1,064 Canadian franchisors with 49,711 franchises.[3] Business-format franchising includes everything from Baskin-Robbins to Merry Maids, from Sheraton Inns to diet centers. Century 21 already claims to be the largest real-estate organization in the world,[4] and Ramada advertises that it is the world's "third largest and growing hotel chain." These are companies which have given, to what frequently are existing individually owned enterprises, the advantages that derive from sharing a recognized name. They also offer the central organization and expertise that was previously exclusive to international chains. Newspaper advertisements offering franchises to the public demonstrate that franchising now covers a very wide range of goods and services.

Like the North American economy in general, business-format franchising tends to be showing the greatest growth in the service industries. This is because some of the manufacturing industry is moving overseas, and its products are frequently returning to be distributed under North American brand names. But the growth in service industries is also to be expected in a country where the human franchise is spreading fast. When the Saddler, in Defoe's *Journal of the Plague Year*, describes his household in 1665, he talks of his "servants." This refers not only to his domestic servants, but also to those who worked in his business and would normally have resided with him.[5] In the seventeenth century, the so-called servant class was the largest of all classes. Servants supplemented the work of married women, who were still regarded as chattels—the word has the same root as cattle. Three hundred years later, though we are the descendants of such people, we certainly do not like to think of ourselves as either servants or chattels. However, much of the same work must still be done, and there is a great deal of it. Hence the remarkable demand for what we euphemistically call our "service industries."

Franchising offers particular benefits to the United States, where it began. This is because the so-called business-format franchising is particularly suited for exporting the American way of life. Matthew Boulton used to say of his steam engines that they offered what all the world wanted—power. Though they may damn them vociferously, what all the world wants today are the material goods and lifestyle associated with North America. An article in *Venture* (July 1985) points out that, in 1971, 156 American companies operated 3,365 franchise outlets overseas. By 1983 that figure had doubled to 305, but the number of franchise outlets had increased to 25,682, and 127 more companies were considering overseas expansion.[6] Doris Walsh reported in 1986 that 50 percent of Coca Cola's profits came from overseas sales. She listed the most popular countries for American overseas franchising as Canada, Japan, the United Kingdom, Australia, France, West Germany, Singapore, and Malaysia, in that order.[7]

From the point of view of the United States, international franchising offers the attraction that it brings in foreign currency in a period when American labor is still finding it difficult to compete in foreign markets. Moreover, this rationalized packaging of the American way of life may well prove to be a more successful form of indoctrination than the export of munitions.

EARLY DEVELOPMENTS IN WESTERN BANKING AND FINANCE

The modern world has required a rationalization not only of retail distribution but also of finance, which now oils the wheels of virtually all human activities. The Reformation gave the northern and Protestant countries of Europe a decisive lead in banking. This was because they were able to come to terms with the usury laws, which had prohibited the lending of money on interest. These laws had been the greatest single factor holding back the development of European money markets in the modern sense. In 1547, at Geneva, John Calvin fixed the maximum rate of interest at 5 percent. In other words, usury would now be acceptable as long as it could be held within bounds. Calvin's action really did bring down the rate of interest, for much the same reason that the price of liquor came down when America's prohibition was repealed.

By 1609 the Bank of Amsterdam had been founded, and it provided credit for such institutions as the Dutch East India Company and the city of Amsterdam. In 1694 the Bank of England was founded by the Scotsman William Paterson in return for a subscription of £1.2 million to be lent to the king at an interest of 8 percent. England's maximum rates of interest—which did not, however, apply to crown loans—had been restricted to 10 percent in 1571–1624; 8 percent in 1624–51; 6 percent in 1651–1714; and 5 percent after 1714. The usury laws were not finally repealed until 1854.[8] In the earlier stages, government loans had been made at or above the usury limits. However, the development of banking institutions—and particularly the Bank of England after 1694—resulted in a sharp increase in government borrowing. Despite this, the institution of banking meant that the interest rates which had to be paid became progressively reduced. Much of the credit for the lower rates must go to the Glorious Revolution of the Whigs in 1689, and perhaps also to the Huguenots, who provided seven of the initial twenty-four directors of the Bank of England. A newsletter of 1686 claimed that Huguenot merchants, anglicized descendants of earlier refugees, and "other merchant dissenters" possessed "six parts in ten of the moving cash that drives the trade of the whole nation." The influence of the Huguenots in England, particularly after the revocation of the Edict of Nantes in 1685, was much greater than the number of immigrants—forty to fifty thousand—might suggest.[9]

In the eighteenth century there was a remarkable increase in monetary as distinct from landed wealth, and the best yardstick is the expansion of the national debt. One millon pounds at 10 percent, floated in 1692, was the first English national debt of long maturity. By the time of the French wars of 1793–1815, the national debt had increased to nine hundred million pounds. There seems to have been an almost symbiotic relationship between the development of bourgeois values and the growth of the national debt. The arrangement was just as attractive to individuals, who wanted to receive regular and reliable interest payments on the increasing amounts of capital that were accumulating, as it was to the government. Defoe was all too well aware of the tastes and needs of his readers. The only explicitly anachronistic element in *Moll Flanders* is a careful explanation of how to deposit money in the Bank of England, which was founded in 1694, even though the manuscript of the novel was ostensibly written in 1683.[10]

By 1770 there were 50 banks in London. We find that by 1776, after the expansion of towns in the North during the eighteenth century, there were about 150, and, by 1800, 350 banks in the provinces. The proprietors included the Quaker bankers, Barclays, who, like many in the banking system, were Nonconformists. After the Restoration of 1660, the Clarendon Codes had permitted only Anglicans to hold positions in the government, civil service, and universities. In trade, however, the Nonconformists prospered rapidly. And, as time produced money for them, some of the hellfire went out of their bellies. I have described elsewhere how the infusion of City cash into the aristocracy by way of dowries, and the retirement of City merchants into the home provinces around London, played an important part in the English economy of the eighteenth century.[11]

As reflected in the novels of both Defoe at the beginning and Austen at the end, 5 percent could be taken as a good average rate of interest in England throughout the eighteenth century. Sidney Homer describes the development of the new money market: "Only after the Revolution of 1689, when King William III and the Parliament could borrow in the name of the united country, and could offer some promise of joint fiscal responsibility, was the English national credit established. Until the

1690s England, although making great commercial progress at home and abroad, had no central bank and no organized money market."[12] In 1749 Henry Pelham was able to persuade investors to convert 54 million pounds of government debt into 3 percent consolidated annuities or "consols." It says much for the strength of the British money markets that throughout the French wars the ten-year average yield on these never rose above 5 percent: 4.64 percent for the 1780s, 4.54 percent for the 1790s, 4.80 percent for 1800–1809, and 4.80 percent for 1810–16. Thereafter, in time of peace, there was a steady fall in the ten-year averages until they reached 3.16 percent for the 1850s. This uniformly low rate of interest reflects a continuing combination of low labor costs and little, if any, inflation.

Low interest rates were conducive to improved demand and production, but so was the bourgeois drive, which derived in no small measure from the desire to acquire gentility after retiring from trade. Defoe, who had a keen personal understanding of this quality, writes with pride: "I dare oblige myself to name five hundred great Estates, within one hundred Miles of *London*, which within eighty Years past, were the possessions of the antient *English* Gentry, which are now bought up, and in the Possession of Citizens and Tradesmen."[13] And Defoe argues, both in his literary and nonliterary works, that a tradesman should retire into the gentry when he has twenty thousand pounds: "Every Man that is advanc'd in Business, so as to be worth twenty thousand Pounds should leave off," because " 'twould be much better for the Nation in general, that twenty Tradesmen, with each a thousand Pound in his Pocket, were employ'd in Trade, than one Tradesman with twenty thousand Pounds" (*Compleat English Tradesman* 2.2:96–98).

The concept of preferring more tradesmen, each with less capital, foreshadows the entrepreneurial strength of the "hungry" franchisee, which is currently being tapped by franchisors. But the idea that a man could and should retire on a capital of twenty thousand pounds, or its interest of one thousand pounds at 5 percent (one hundred thousand and five thousand dollars respectively in U.S. currency), clearly belongs to a period earlier than our own. Yet, in fact, such an income continued to be more than satisfactory in both Britain and the United States for no less than 250 years after the founding of the Bank of England. It is only since World War II that the demands for higher wages and the concomitant reduction in productivity have resulted in a level of inflation not experienced in those countries since England's Whig revolution of 1689. When people live beyond their means they go into personal bankruptcy. When nations do so—as most in the Western world are doing now—the value of their money goes down. As a result they are often able to repay debts with cash of lesser value, but the impact on the middle class and the elderly can be catastrophic. South America's present condition is a reflection of such fiscal irresponsibility taken to extremes, and the rise of Hitler hinged largely on the response of the German middle class to the acute inflation that they experienced in 1923.

SOME CONSEQUENCES OF DEMOCRATIZING
THE OWNERSHIP OF WESTERN MONEY

Until after World War II, government spending—which often overheats the markets and drives up interest rates—derived extensively from the purchase of the munitions of war. This expenditure has, however, now been equaled and even exceeded by disbursements related to social services. Such expenditures derive directly from the spreading of democracy and the franchise: the rationalization of people themselves. Unlike the earlier period in the rise of Western technology, none of us now feel that we should be looking forward to the workhouse as the place in which to spend our latter days. With the help of government tax subsidies and supplements from our employers, many of us accumulate even more funds than we had anticipated. In the United States there were $1.8 trillion in pension funds by 1988. In the absence of the inevitable "inflation tax," such funds would provide us with quite luxurious retirement pensions. Thanks to Western technology, the descendants of peasants have never been so comfortably endowed. Though most of us know very little about investment, there are always friendly mutual-fund and life-insurance salesmen, together with franchised financial advisers, who will help us to manage our money.

The upshot is that there are now hundreds of billions of dollars being traded in and out of the markets by people who do not own them. An increasing proportion of this trading is fueled by the expectations of salesmen, who incorporate inflationary capital gains as though they were part of the interest earned by their mutual funds. We might do well to remind ourselves of what inflation has done to the American stock markets. In the period 1906–24, the Dow Jones Industrial Average traded up from 100 to just under 125. It jumped to 381.17 in 1929 and then came down rather quickly to 41.22. Between 1937 and 1950 it traded between approximately 93 and 212. Then the first real rise took place. Nevertheless, the Dow still traded between about 535 and 1,000 for the twenty-five years from 1957 to 1982. Between 1982 and the "October massacre" of 1987 (the first crash on our global financial markets), the Dow moved directly upward in an unprecedented rise from 1,000 to about 2,700. Needless to say, it was not at all difficult for mutual funds—which make a practice of incorporating inflation-driven capital gains in their so-called earned interest—to show a good profit during that period. Since most of us are addicted to anticipating future earnings from past profits, our own greed greatly helps the managers of such funds.

More recently, money markets have been overheated by yet another destabilizing factor related to democratization. When the London Stock Exchange grew out of Jonathan's coffeehouse, at the beginning of the period with which we are concerned, most businesses were run and owned by their proprietors. The South Sea Bubble of 1720 should have acted as a warning that stock exchanges can be dangerous places for widows, orphans, and the unwary. We have seen that it was not until the sons of the original proprietors took over the management of Boulton and Watt that the idea of employing professional management began to develop in the nineteenth century. But even in that century, companies tended to be owned or at least controlled by the founder or his family. The directors of a company generally had a major investment in its shares and could be relied upon to control the management. As a result, the management could normally be expected to operate in the interests of the owners.

In our own century, with the development of larger and larger companies and the spread of the franchise, there has been a tendency for stocks to be more and more widely dispersed. In the United States, perhaps one million people normally held shares in the years before World War II. The single exception was the period from 1926 to 1929, leading up to the stock-market crash, when there were between four and five million investors. But Ford's announcement that it was going public in 1955, was, in effect, an invitation to three hundred thousand new stockholders. Such figures now seem puny. In 1986–87 a survey commissioned by the New York Stock Exchange showed that around forty-seven million Americans owned shares, an increase of five million over 1983.

At the end of 1986 Philip Beresford and John Cassidy reported that Margaret Thatcher's privatization policy was producing similar trends in Britain. Britain's privatization has ensured that small numbers of shares would be issued to as many potential capitalists as possible. At that time the proportion of the population owning shares in Sweden and Norway was 25 percent, in America 20 percent, in Britain, Japan, and France 16 percent, and in Belgium 12 percent.[14] In the absence of a major recession this upward trend in the ownership of shares will certainly continue.

On the face of it, one should rejoice that capitalism is clearly becoming democratized. But for the cynics there is another side to the coin. The first and perhaps not so questionable implication is that widespread shareholding tends to boost the sales of almost any company's products. What is perhaps more sinister is the way in which small shareholders are now buying stocks either directly or indirectly through large institutional investors, like mutual and pension funds. This increasingly ensures that there is no effective control of the company by those who own the shares. The result of the democratization of shareholding—in contradistinction to the earlier proprietors of the nineteenth century—seems to have received relatively little attention. And well it might not from the present directors and managers of large corporations. In many cases they have become a self-perpetuating collection of people with very little of their own capital invested in the organizations for which they are responsible.

Many of today's corporate raiders, greenmailers, and specialists in leveraged buyouts and takeovers thrive on the conditions that have just been described. When the management of a company cannot be controlled by the shareholders, it is bound to give a high priority to its own incomes and even to its own golden handshakes. If such a company seeks a white knight in order to avoid a takeover, the management can generally be relied upon to protect its own interests rather than those of the shareholders. The raider, for his part, will generally use other people's money to seed his initial purchase of shares. And it is remarkable with how much less than 50 percent of its shares a widely held company can frequently be controlled. The raider will often use "junk" bonds—those offering high interest but low collateral—to make the purchase. He will then frequently employ the proceeds from selling all or part of the assets of the purchased company to pay for the cost of taking it over.

It would be nice to think that, though red in tooth and claw, company takeovers were part of a Darwinian process that did at least reinvigorate industry by ensuring the survival of the fittest. In fact, apart from weeding out some inefficient managers, this widespread financial activity does very little to improve the productivity of the nation. What it does suggest is that—with the exception of certain areas like public utilities, transport systems, hospitals, education authorities, state and municipal services, and perhaps banks—large corporations may not now be as much of an asset to a nation as they were in the past. If subcontracting and franchising were permitted to take place with as little hindrance as possible, an economy based extensively on small businesses would function far better than we generally imagine. What Alexander Pope called "SELF-LOVE," in his *Essay on Man*, is a much more potent catalyst for turning an entrepreneur's time into money than we tend to recognize.

RATIONALIZING GOVERNMENT, CORPORATE, AND PERSONAL FINANCING IN A GLOBAL VILLAGE

North America does not operate in a vacuum. In finance, as in much else, the world is now becoming a global village. At the close of the seventeenth century, Britain took over financial leadership from the Dutch, and after World War I the United States took over from Britain. Similarly, Japan, in the middle of the 1980s, began to take over from the United States. As John Kohut reports, at the beginning of the 1980s the net foreign assets of the United States were worth about $140 billion (U.S.) while those of Japan were worth about $11 billion. But the luxury of a continuing trade deficit is bound to show up on the books. Halfway through the eighties, the United States (for the first time during the lives of most people) had become a net debtor nation. Japan, by then, had dramatically changed its role. It was now the world's largest creditor, with net external assets of about $130 billion. As recently as 1983 the United States was still the world's largest creditor nation with an investment cushion of $89.4 billion, but by 1987 its debt of $368.2 billion had made it by far the world's largest debtor nation.[15]

The increasing value of the yen has meant that by April 1987 the value of the companies traded in Japan's stock market exceeded that of the United States, for the first time. The increased value of the yen has also meant that Japanese workers, though their living standards might not yet reflect this, are now the highest paid workforce in the world. The Japanese save at about four times the rate of the Americans. Since, in addition, they have large trade surpluses—nearly $83 billion in 1986—their financial institutions, including pension funds, have very considerable sums to invest. A part of these sums is now being invested abroad, and particularly in North America. In the United States, Japanese investment in stocks and bonds was $40.1 billion in 1985, with an estimated increase of 50 percent for the following year.[16] In effect, Japanese savings have been helping to finance American deficits. There have also been large Japanese investments in American real estate, as well as in automobile and other manufacturing plants. The Japanese now feel that the rise in the value of the yen will make their own labor costs too high.

There seems to be a remarkable myth propounded by economists regarding the value of a falling dollar. It was earlier heard in Canada, when the value of the Canadian dollar fell from $1.06 U.S. to around $.75 U.S. Apparently this was much to be desired, because it meant that Canadian products might therefore be more easily exported to the United States. What was ignored was the fact that Canadian labor and management should have been more productive in the first place, and that management should not have agreed to excessive wages and benefits for the purpose of avoiding strikes. With tighter and leaner production methods, the Canadian dollar would not have needed to fall in order to discount the value of excessive wages.

Since 1985 United States economists have been echoing the earlier statements of their Canadian colleagues and praising the advantages in terms of the exports that might be expected from a falling American dollar. As Halberstam puts it, with regard to the American automobile industry: "After an ugly strike or two the postwar managements of the different companies had evolved a policy of bringing labor in, making it in effect a junior partner, granting much of its wage demands and passing on the additional labor costs to the customer."[17] Today, the total cost of this cozy arrangement at the public's expense is threefold: the Americans have lost considerable market share to the Japanese; their government has a large and increasing deficit; and the fall of the U.S. dollar means that all wages, including those of union workers and management, have dropped by at least 40 percent on the international market. The Japanese—who are the latter-day equivalents of the British Nonconformists—are, like other foreigners, now finding that more and more of their auto parts are being sold under American hoods. In addition, American auto companies, like many other American organizations, are using their own brand names to distribute Japanese and other Pacific Rim manufactures.

It would be a mistake to underestimate the financial strength of the United States. Although it currently maintains prohibitive annual deficits, it still had the greatest amount of foreign assets in 1985. At that time, the Bank of England estimated that the United States had $952 billion, Britain $859 billion, Japan $440 billion, and the Arab oil producers $438 billion. However, Japan appears to be exporting capital at the rate of about $150 billion a year. In a smaller way, South Korea, Taiwan, Hong Kong, and Singapore may soon be following suit. Though the Arab billions swish around the world's money markets, any increase in their stake will depend upon the price of oil. Neither the Arabs nor the Japanese are anxious to lend money to Third World countries; and Britain, Canada, and the United States may prove to be somewhat less forthcoming after the experience with what their bankers euphemistically call their "sovereign loans."

There is a lot of money swilling around in the global village, perhaps more than is good for the system. Einstein once said that to ask the right question is already to be a good part of the way toward finding the right answer. The developments that we have been following in this study would suggest that our question might be: If we are moving toward a global village, what can be done to rationalize further the financial system? The latter part of the 1980s has seen the introduction of twenty-four-hour stock-exchange and currency trading between the main markets of the world, using complex computerized installations. This is a step in the direction of rationalizing the financial systems of the world. It strongly suggests that there will also be a related demand for a rationalized currency system.

We have seen, thus far, that Western technology has demanded the rationalization of production, transport, and distribution. And we have observed that this has been considerably facilitated by the rationalization of time, calendar, and chronology, as well as the development of the decimal and metric systems and the English language. In all these areas, Western civilization has, in some measure, succeeded. However, the rationalization of a global currency still seems to remain an elusive dream. In 1986 the non-Communist world's gold production was 1,280 metric tons, in addition to 402 metric tons exported by Russia and China. This hardly suggests a stock large enough to back a global currency for five billion people. Neither does the international trust necessary for a global paper currency seem possible at present.

In individual Western countries we have generally been able to maintain a respect for currency by providing a greater franchise and equality between people than we tend to recognize. This means that we normally do not hide gold under the mattress, but also that bankers expect to be supervised by a government which represents ourselves. The trust in financial institutions is complemented by the fact that, in individual Western countries, people who want to improve their material welfare can move to areas where there is a shortage of labor. As this study has shown, the melting pot and the ability to migrate were essential to Western technology. Furthermore, Western technology has both created and benefited from the increasing franchise of human beings. However rocky the road, and despite our best efforts to the contrary, Western technology may yet ultimately impose upon all of us what we most need. This includes the franchise, the right to migrate, the curbing of our population growth, and the relative equalization of all human beings. If and when such a mundane miracle materializes, the problems of rationalizing both currency and sovereignty will have been virtually solved. They have long been more closely interwoven than we may have wished to recognize.

IV Rationalizing Human Beings

X Labor and the Individual

When we considered the application of time and method study to manufactures, we did so from the point of view of rationalizing production. Now we must consider rationalization from the somewhat more personal perspective of its effect on men and women. Let us remind ourselves of the inaccuracy of measurements in the period before Britain's Restoration of 1660. In a chapter dealing with "a general computation of men, and cattel's labours: what each may do without hurt daily," G. Markham had the following to say, in 1660:

> A man . . . may mow of Corn, as Barley and Oats, if it be thick, loggy and beaten down to the earth, making fair work, and not cutting off the heads of the ears, and leaving the straw still growing one acre and a half in a day; but if it be good thick and fair standing corn, then he may mow two acres, or two acres and a half in a day; but if the corn be short and thin, then he may mow three, and sometimes four Acres in a day, and not be overlaboured.[1]

The day was still the statute dawn-to-dusk working day, defined as late as 1725 as "from five in the morning till betwixt seven and eight at the night, from the midst of March to the middle of September."[2]

The hours to which Markham refers were, of course, later transferred to the factory. There, however, they applied throughout the year and were imposed under much more closely controlled conditions than had been set out for work on the farm. We have a tendency to conjure up a golden age of the past, when men and women who worked on the land enjoyed a life of bucolic simplicity. Stephen Duck (1705–56) was one of the few poets who could write of the "master's curse" from the laborer's point of view. Duck describes a typically oppressive farmer in *The Thresher's Labour*:

> He counts the bushels, counts how much a day,
> Then swears we've idled half our time away.
> "Why look ye, rogues! D'ye think that this will do?
> Your neighbours thresh as much again as you."[3]

THE RATIONALIZATION OF LABOR IN AN URBAN SETTING

Until the division of labor permitted the simplification of work, factory owners were still obliged to encourage the acquisition of skills. In the factory of Boulton and Watt, "Fathers were induced to bring up their sons at the same bench with themselves, and initiate them in the dexterity which they had acquired by experience; and at Soho it was not unusual for the same precise line of work to be followed by members of the same family for three generations."[4] In the pottery factory of Josiah Wedgwood (1730–95)—just as in clock and other factories in the latter half of the eighteenth century—there was an ongoing movement toward the simplification of work. Simpler jobs, like painting the bordering on china, were given to women. In Wedgwood's own words, he set himself twin tasks, first "to make *Artists* . . . [of] mere *men*" and, second, to "make such *machines* of the *Men* as cannot err."[5]

Writing in 1835, Andrew Ure suggests that employers were anxious to extend as far as possible the subdivision of labor, in order to diminish the power of potentially recalcitrant skilled male workers. They also wanted to circumvent the incipient unions by introducing as much machinery as possible: "By the infirmity of human nature it happens, that the more skilful the workman, the more self-willed and intractable he is apt to become, and, of course, the less fit a component of a mechanical system, in which, by occasional irregularities, he may do great damage to the whole."[6] Among "occasional irregularities" one may include the widespread practice on the part of skilled workmen of taking off Saint Monday (and sometimes Tuesday and Wednesday too), to the detriment of the employer, as well as the women and children who depended on the skilled craftsman for their work.[7] Instead, Ure advocates "the system of decomposing a process into its constituents [subdivision of labor], and embodying each part in an automatic machine." As he argues, "on the automatic plan, skilled labour gets progressively superseded, and will, eventually, be replaced by mere overlookers of machines." This, he notes with some foresight, will result in "the equalization of labour."[8]

Ure's predictions are, however, not as well grounded when he argues that machinery will overcome the potential for unions. After one and a half centuries, what he has to say about unions does, nevertheless, make interesting reading:

> It is one of the most important truths resulting from the analysis of manufacturing industry, that unions are conspiracies of workmen against the interests of their own order, and never fail to end in the suicide of the body corporate which forms them; an event the more speedy, the more coercive or the better organized the union is. The very name of union makes capital restive, and puts ingenuity on the alert to defeat its objects.[9]

The "ingenuity" that Ure talks about resulted in the yoking of people to the speed set by machines. Sir James Kay-Shuttleworth describes the direct relationship between people and machines: "Whilst the engine runs the people must work—men, women, and children are yoked together with iron and steam. The animal machine—breakable in the best case, subject to a thousand sources of suffering—is chained fast to the iron machine, which knows no suffering and no weariness."[10]

Charles Babbage tells a revealing story about the boom in machine-made lace in the Nottingham area during the twenty years leading up to 1830. He describes the painful attempts by those working with inferior technology to keep up with those who could afford the increasingly more efficient machinery. The trade employed "above two hundred thousand," and, for making the new machinery, "those who were best paid, were generally clock and watch makers, from all the district round."[11] Thus watchmakers were not only providing accurate time, but they were also contributing widely to the manufacture and maintenance of machinery for the British industrial revolution.[12] Machinery rationalized and regulated not only the labor that was tied to it but also the sweated labor of those who had to try and compete with it (figure 8).

During this same period, in the early part of the nineteenth century,

Figure 8. This photograph of the girls' dining room of the H. J. Heinz Company illustrates a typical uniformity among factory employees. (From *Wage Earning Pittsburgh*, ed. Paul Underwood Kellogg, vol. 6 [New York: Survey Associates, Inc., 1914]; courtesy of McPherson Library, University of Victoria)

what came to be known as the American System of Manufacture went one step further than the British industrial revolution. Now the machines, rather than their operators, became specialized. This allowed the integration into the factory system of the unskilled labor that came to America from lands with no industrial background. The development of the American System of Manufacture is credited to Eli Whitney, who first instituted mass-production methods for making muskets. By 1822 Eli Terry, probably influenced by Whitney's methods, was producing six thousand clocks a year and selling them at fifteen dollars each. Whitney himself turned to clockmaking after 1848. Soon enough, locomotives, sewing machines, bicycles, motor cars, typewriters, and much else were being made by using the interchangeable parts that were associated with the American System of Manufacture.

A further step required for rationalizing the time spent by operatives in factories was the introduction of artificial light. This was first effected by William Murdock, who adapted the Argand burner. In

1798 he installed gas lighting in a factory near Birmingham, England, and was soon providing this facility commercially for several institutions in the neighborhood. The use of artificial lighting was an essential prerequisite for developing a fully rationalized factory. James Sexton demonstrates well how Aldous Huxley's description in chapter 11 of *Brave New World* (1931), dealing with rationalized workers in a rationalized factory, parodies with considerable acuity what Ford and Taylor were undertaking in real life.[13]

In the early days of the British industrial revolution, the recalcitrance and lack of flexibility that employers claimed to find in skilled male labor had made them turn to cheap child and female labor in numbers not previously experienced. The following letter from Lewis Paul to John Russell, fourth duke of Bedford, is reputed to have been drafted by none other than Samuel Johnson: "I therefore take the liberty of proposing to your Grace's notice a machine (for spinning cotton) of which I am the inventor and proprietor, as proper to be erected in the Foundling Hospital, its structure and operation being such that a mixed number of children from five to fourteen years may be enabled by it to earn their food and clothing."[14] Johnson had drafted this letter about 1758. Almost a century later, writing of his childhood in the Yorkshire of the 1840s, George Oldfield recounts:

> My eldest sister went to work in the factory very early. I soon had to follow, I think about 9 years of age. What with hunger and hard usage I bitterly got it burned into me—I believe it will stay while life shall last. We had to be up at 5 in the morning to get to the factory, ready to begin work at 6, then work while 8, when we stopped ½ an hour for breakfast, then work to 12 noon; for dinner we had 1 hour, then work while 4. We then had ½ an hour for tee, and tee if anything was left, then commenced work again on to 8.30. If any time during the day had been lost, we had to work while 9 o'clock, and so on every night till it was all made up.[15]

From another point of view, here is a contemporary impression of Manchester provided by Anthony Ashley Cooper, seventh earl of Shaftesbury: "What a place is Manchester—silent and solemn; the rumble of carriages and groaning of mills, but few voices and no merriment.... Thirty-five thousand children, under 13 years of age, many not exceeding 5 or 6, are working, at times, for 14 or 15 hours a day, and also, but not in these works, during the night!"[16] But it was no better in the countryside: "Two shillings a week for lads twelve and fourteen years old.... And by those youths and young men two-thirds of all the ploughing and carting of the farm is done. They are hired from a distance in almost all cases; are hired by the year; provide themselves with food and clothing out of their wages; sleep in a stable-loft or barn, having no fireside to go to; no hot dinners, but everlasting bread and lard, bread and lard, bread and lard!"[17] In 1868 Elihu Burritt wrote that "a writer, who visited the different brick-making establishments of the [Black Country], estimates that seventy-five per cent. of the persons employed are females; and perhaps two-thirds of these are young girls from nine to twelve years of age."[18]

It is easy for us to be shocked by many of the values embodied in the industrial revolution, but it is important to remember that Western technology grew out of times that were very little upset by the idea of slavery. John Locke, the Whig philosopher, approved of it; Daniel Defoe's only caveat was that a modicum of humanity on plantations would enhance the productivity; and the Anglican church condoned slavery in much the same way that, in our own century, the Lutheran church simply ignored the Holocaust. It is one of the ironies of history, which has never been satisfactorily explained, that the sentimental movement (in contradistinction to what Allardyce Nicoll calls "aristocratic cynicism") is first reflected in literature, in the sentimental plays of the late seventeenth century, and in Laurence Sterne's *Sentimental Journey* of 1768. The sentimental movement is every bit as much an offspring of the bourgeoisie as is the enslavement to time of themselves and their employees since the industrial revolution.

SLAVE LABOR AND WAGE LABOR

Sir William Petty (1623–87)—a pioneer of economic thought and economic and demographic statistics—makes a recommendation that combines contemporary attitudes toward both child labor and slavery. His proposal (which demonstrates that Jonathan Swift's *Modest Proposal* [1729] was not ungrounded in contemporary thought) offers a solution for the difficulties experienced by poor pregnant women: "Lett the Government in humanity make provision for every woman with Child for 30 days, the woman leaving her Child to be a servant to the

Government for 25 yeares, suppressing the names of their parents. The Charge of maintaining a woman 30 dayes in Child bed may well be defrayed for under 30 shillings; but if the value of mankind be in this age & country 70 pounds per head, a new born Child, bread up to fair work for 25 yeares, will be very well worth 3 times 30 shillings, as may be seen in the price of Negros [sic] Children in the American plantations."[19]

Petty's valuation of human beings at "70 pounds per head" might offend our modern sensibilities, but it certainly reflects his age. James Burke indicates that under the British manorial system, toward the end of the Dark Ages, a person's worth was determined by his value when murdered. The penalty was either death or the payment of *wergild* (the body price of the victim): 1,200s. for a noble, 300 for a thane, 200 for a churl or free peasant, but nothing for a serf. Needless to say, a woman was worth less than a man of equal social standing.[20] During World War II there was an unofficial body price in India and Ceylon for the accidental killing of a native by British troops. This was 600 rupees, or about $120. We still have such values, though they are much higher now. Because our sentimental indoctrination no longer allows us to state them openly, lawyers' "percentage fees" take between 30 and 40 percent of most "body-money" claimed in Western litigation, and drive up insurance rates accordingly. The fact that there are more lawyers in the New York area than the whole of Japan should tell us something about the counterproductive litigiousness of our society.

The eighteenth century had other evils. The slave trade then supported some of England's greatest families, not to mention the plantation owners in America. What is known as the British triangular trade involved, first, the purchase of slaves through local traders on the west coast of Africa; secondly, their exchange for sugar from the Caribbean and tobacco from America; and, finally, the return journey to England. Though hundreds of thousands of African men and women were lost on the Middle Passage, the profits seem to have blinded some very fine families to the suffering that they caused. In *A Plan for the English Commerce* (1728), Defoe argues that slaves must be included in British exports, because they "are the Produce of British Commerce in their *African* Factories." The trade is particularly profitable, since "these Negroes do not cost in the Country above 30 to 50s per Head," and the selling price "is of late Years risen in all the Colonies . . . from

20 to 30 l. a Head, according to the Age the Growth and the sex of the Negroes." Defoe estimates that traders "including the *Assiento* carry 40 to 50000 Slaves a year from the Coast of Africa."[21]

But African slaves were not the only form of cheap labor, and English workers were becoming better indoctrinated with time discipline. E. P. Thompson, in "Time, Work Discipline and Industrial Capitalism," shows how, throughout the eighteenth century, charity schools and factory owners, like Crowley and Wedgwood, indoctrinated their workforces with a new sense of time discipline.[22] Essentially, of course, such workers were "wage slaves," who had very few options for using their theoretical freedom. Robert Southey, the Romantic poet, makes a direct comparison between factory workers and Negro slaves, when visiting the Lanark mills of Robert Owen. He does this even though, in their own day, Owen's cotton mills were considered to be Utopian:

> Some 200 children, from four years of age till ten, entered the room . . . performing manoeuvres the object of which was not very clear, with perfect regularity. . . . I could not but think that these puppet-like motions might, with a little ingenuity, have been produced by the great water-wheel, which is the *primum mobile* of the whole Cotton-Mills. . . . Owen in reality deceives himself. He is part-owner and sole Director of a large establishment, differing more in accidents than in essence from a plantation: the persons under him happen to be white, and are at liberty by law to quit his service, but while they remain in it they are as much under his absolute management as so many negro-slaves. His humour, his vanity . . . lead him to make these *human machines* as he calls them (and literally believes them to be) as happy as he can, and to make a display of their happiness.[23]

It goes without saying that religious education was invariably part of such arrangements. Joseph Farington's diary for 1801 records his pleasure at seeing children of ten or eleven years of age leaving Sir Richard Arkwright's factories in Cromford after working for thirteen hours. He gives "great credit" to Arkwright for building a chapel. The boys go there on one Sunday, and the girls on another. On the alternate Sundays, they receive instruction from "an Old Man their School Master."[24]

Western technology did not develop painlessly. It was motivated by greed, and there was little if any succor for those who fell by the wayside (figure 9). Though the population turned frequently to church and

Figure 9. Gustave Doré's *Wentworth Street, Whitechapel* suggests that there was plenty of cheap labor available in nineteenth-century London. (From *London: A Pilgrimage*, by Gustave Doré and Blanchard Jerrold [London: Grant, 1872]; courtesy of McPherson Library, University of Victoria)

chapel for solace, the dominating concern was with economic survival in this world. In such a Darwinian struggle, conceived as the survival of the fittest, the new type of knight would be modeled after Sir Richard Arkwright (1732–92), rather than Galahad or Launcelot. Arkwright was born in Preston, Lancashire, the virtually uneducated youngest of seven children, and brought up to be a barber and wigmaker. Our history books tell us that Arkwright's water frame and Hargreave's spinning jenny are the key inventions of the textile industry, which spearheaded the British industrial revolution. Both came into use in 1768. Arkwright patented his water frame in 1769 and a number of minor inventions in 1775. At a trial in 1785 judgment was, however, given against these patents. It turned out that Highs and Arkwright both claimed to have invented water frames that had almost certainly been made in the first place by the Warrington clockmaker John Kay. Of related interest, and pointing to the close relationship between clockmaking and the industrial revolution, is the fact that the mechanism of Arkright's frame was known as "clockwork." Also, in his first patent he had falsely described himself as "Richard Arkwright . . . clockmaker."[25]

Arkwright was, however, a survivor. By 1782 he had a capital of two hundred thousand pounds and employed five thousand workers. This was the sort of man whom the nation rejoiced to honor, and four years later he was knighted. And such, too, have been the men—like the Rockefellers, Carnegies, Fords, and an assorted collection of railway barons—whom the Americans have more recently rejoiced to honor. In England, however, the large organizations built up by men like Crowley, Wedgwood, and Arkwright—and reflected in literature by the Bounderby of Charles Dickens's *Hard Times*—engendered a social discord that has still not been resolved. The privileges of the landed aristocracy, though probably no more deserved, seemed to have been sanctified for the peasants on the basis of tradition. But the already displaced urban workers appeared to consider the new form of bourgeois power more brutal and more naked.

HUMAN LABOR AND THE UNION MOVEMENT

By 1779, in a letter from Josiah Wedgwood to Thomas Bentley, we get an early image of "the mob" attacking mills. According to Wedgwood, the owner of a mill near Chorley was able "to repulse the enemy and preserve the mill for that time. Two of the mob were shot dead upon the spot, one drowned, and several wounded. The mob had no firearms. . . . They were greatly exasperated, and vowed revenge." As a result, they collected firearms, melted down their pewter dishes into bullets, and were joined by the duke of Bridgwater's colliers and others to the number of some eight thousand men. On Monday "the mob completely destroyed a set of mills valued at 10,000 [pounds]." They then marched on to Bolton, "and their professed design was to take Bolton, Manchester, and Stockport in their way to Crumford, and to destroy all the engines, not only on [sic] these places, but throughout all England."[26] This typifies the response to rationalization by both employers and employees. Over the years it has developed into what is now known as "the British disease."

Sixty years after Josiah Wedgwood's letter, the activities of the Chartists demonstrate that all that had changed was the size of the mob. General Sir Charles Napier describes the situation in 1839. He does so from the more neutral viewpoint of someone who is responsible for restoring peace: "This district could easily turn out three hundred thousand people on Kersall Moor; and the Chartist newspapers asserted that they would turn out five hundred thousand. I did not believe this, but secretly thought one hundred thousand might be assembled:—quite enough to render my position very dangerous. My two thousand men and four guns were indeed enough, if well handled, but not enough to afford mistakes." Napier complains that neither the Whigs nor the Tories are prepared to come to some form of conciliation with the workers, whose own employers and leaders govern them ill enough: "Good God what work! to send grape-shot from four guns into a helpless mass of fellow-citizens; sweeping the streets with fire and charging with cavalry, destroying poor people . . . reduced to such straits that they seek redress by arms, ignorant that of all ways that is the most certain to increase the evils they complain of."[27]

Communism, as a religion of the new urban proletariat, grew out of the British hotbed of industrial experience, exacerbated by the density of population, rather than out of Russia or China. Friedrich Engels, in *The Condition of the Working Class in England in 1844*, refers to "South Lancashire with its central city Manchester" as the "classic soil on which English manufacture has achieved its master-work." For Friedrich Engels and Karl Marx, the rationalization of manufacturing will not be complete until the urban proletariat take control of their own destinies: "The degradation to which the application of steam-power, machinery, and the division of labour reduce the working-man, and the attempts of the proletariat to rise above this abasement, must likewise be carried to the highest point and with the fullest consciousness."[28] Though the proponents of the Marxian dialectic are clearly in favor of a further rationalization of human beings, Communism, in practice, has succeeded only in replacing one set of opportunists with another.

In 1831 some of the secret oaths sworn by the members of a union club were reported in the newspaper *Cambrian*. They include the following, which give an insight into the developing proletarian mentality:

> 3. I will never instruct any person into the art of coal mining, tunnelling, or boring, or engineering, or any other department of my work, except to an obliged brother or brothers, or an apprentice.   So help me God.
> 4. I will never work any work where an obliged brother has been unjustly enforced off for standing up for his price, or in defence of his trade.   So help me God.
> 11. I will never make these obligations known to either master, manager, or underkeeper, overlooker, book-keeper, or any person, except a legal obligated brother.   So help me God.[29]

Poor God, who is probably tired of fighting on both sides during so many bloody Christian wars, was now being invoked in yet another internecine struggle. It was destined to prove every bit as bitter as the others.

Though union battles have certainly been as large and often as fierce in America, they do not seem to have indoctrinated employers and employees with quite the same antagonistic attitude as has occurred in Britain. Perhaps the size of the United States, together with its newness and potential for upward mobility, have helped to dissipate some of the bitterness. But bitterness there certainly has been. As in England, the worst was in the largest northern units of the manufacturing industry. Detroit and the United Auto Workers will again serve to highlight the conflict. By the 1930s the auto industry was employing, directly and indirectly, perhaps half a million men, none of whom belonged to a union. David Halberstam reports this as the cruelest type of industrial setting. In the auto industry, the pay, once considered good, was now poor, and the workers were up against brutal foremen, speeded-up production lines, and no protection in hard times. The fight against the auto companies was a hard one. In 1934 General Motors had made $94.9 million, but the average worker earned only $1,100. Yet Franklin D. Roosevelt had said that if he were a working man he would join a union. With the New Deal in progress, things began to turn in favor of the unions.[30]

In February 1937 General Motors capitulated and finally recognized the United Auto Workers. The old era had ended, though Ford held out for another four years. In 1941 Henry Ford signed the first union shop and checkoff contract in the automotive industry. Though Walter Reuther, of the United Auto Workers, tried to avoid passing on the extra labor costs to the public, General Motors was able to insist, at the negotiations in 1946, that this was the company's responsibility. Thus Taylor's prediction that labor and management would be involved in a collusion at the expense of the public had taken place, and would frequently do so again.

For a surprisingly long while, the extent of the collusion between union and management was not apparent. This was because car manufacture was in effect a monopoly. Even Henry Kaiser had failed to break into the industry in the United States, and there was still little enough competition from abroad. But eventually the cozy relationship between management and unions resulted in levels of quality and service unacceptable even to patriotic Americans. As a result of union contracts, labor costs were now indexed to inflation. At the beginning of the seventies, auto and steel workers were being paid 30 percent more than other industrial workers. By the end of that decade, however—with inflation rising as a result of OPEC's success in boosting the price of oil—auto workers were making 60 percent and steel workers 70 percent more than other American industrial workers. The Japanese, who combined low prices with high levels of quality and service, could hardly ask for better conditions in which to break into the American market with their small, reliable, and conservatively designed models.

The Americans have long understood the need for enacting effective antitrust laws against large corporations. John Davison Rockefeller (1839–1937) virtually invented the modern corporation singlehanded. By 1882 his Standard Oil trust controlled 95 percent of the oil-refining business in the United States, in addition to much else. In 1899 this trust was held to be in violation of the Sherman Antitrust Act by the Supreme Court of Ohio. By 1911 even the holding company, a device employed by Standard Oil of New Jersey, was declared illegal by the United States Supreme Court. As a result, this monopoly was forcibly broken up into many independent companies. More recently, we have witnessed a comparable breakup of the property of the American Telephone and Telegraph Company, which was the heir of the Bell Telephone Company, founded in 1877. However, the Americans seem never to have produced parallel antitrust laws against the large unions that were equally capable of dominating a vital area of the economy. In Japan, by way of contrast, labor unions have tended to be restricted to a single large corporation, whose economic health they are therefore very much obliged to consider.[31]

Two aspects of American production during the postwar years highlight the problem. For a short time, in the 1950s, I included Timex watches among the lines that I was carrying as a wholesaler in the Channel Islands. This was before Timex improved quality control and distributed direct to retailers. The shopkeepers to whom I had sold Timex watches were appalled to find that almost 20 percent were being returned by purchasers. The company was all too ready to replace a faulty product, but the public, quite understandably, rebelled at being treated as the first line of quality control. Much the same was occurring in the American auto industry. Yet the world expert on quality control, in whose name an annual award has been given by the Japanese since 1951, is an American, W. Edward Deming. This is clearly a case of a prophet who has not been honored in his own country because his methods are better suited to workers who take pride in their product and their company.

After quality control, the most admired aspect of Japanese production seems to be the JIT or just-in-time system. The system involves bringing auto parts to the assembly line in a synchronized way that avoids costly build-ups of inventory. Precisely this system, though without the name, had existed long ago at Ford's River Rouge plant. But such a system can function properly only when there is cooperation and trust between workers and management. In 1832 Sir James Kay-Shuttleworth, citing Charles Babbage, had already pointed up the problem in England: "If an establishment consist of several branches . . . as, for instance, of iron mines, blast furnaces, and a colliery . . . it becomes necessary to keep on hand a larger stock of materials than would otherwise be required, if it were certain that no combinations [of striking workmen] would arise. The proprietors of one establishment in the trade ... think it expedient always to keep above ground, a supply of coal, for six months."[32] An Associated Press report from Detroit, of November 21, 1986, indicated that "major U.S. automakers have been moving to the just-in-time system, long used in Japan." However, when 7,700 employees walked off "to resolve a dispute over subcontracting jobs and the transfer of some work to Mexico," it "took little more than 24 hours for the strike to affect plants across the United States as G.M. began shutting down assembly lines for lack of parts and sending home thousands of workers, some indefinitely."[33]

Clearly, what is a source of strength in Japan can be used by workers in the United States as the equivalent of hostage taking. But the hostages in this instance are their company, their country, and ultimately themselves. The large companies and the large unions are becoming a considerable impediment to an otherwise remarkably resilient and dynamic American economy. So is the paranoia about communism, which forces Americans to underwrite questionable politicians and economies abroad. It also obliges them to maintain a high-cost military umbrella at home that is based on unionized companies and their bloated managements.

The high price of the yen means that it is no longer even true that Hondas, Datsuns, and Toyotas are flooding the American market because Japanese workers are underpaid. In fact, for some time now, the Japanese have been trying to establish factories in the United States. As Halberstam points out, however, the Nissan people chose Smyrna in Tennessee not just because of the advantages offered by the state or because of its rail connections. A further and crucial factor involved the state's right-to-work laws, and the hope that it might prove far enough from the industrial North to be out of the reach of the United Auto Workers. The experiences of Nissan and other Japanese companies, which have moved to the South, have shown what a more modestly paid management, not driven by a need for immediate results, can do with American labor. In fact, such managements (acting more modestly in their own puritanical tradition) have found that, in the absence of UAW employees, they can still gain pleasure from both giving and receiving a fair day's work for a fair day's pay.[34]

Even the Japanese seem to recoil in terror at the thought of setting up factories in the United Kingdom. The so-called "British disease" seems to permit management and labor to destroy a company rather than come to terms with one another. This is something that, happily for themselves, the Japanese find it difficult to understand. More recently—after a trip to Japan in 1983 by Margaret Thatcher and an announcement of intent in 1985—Nissan has decided to attempt a cutdown version of an auto plant, by way of trying out the adaptability of British workers.[35]

The message could hardly be more clear: in an economy with increasingly world-wide ramifications, neither management nor workers can hold their nations hostage indefinitely. Nor can they expect to draw on the profits of rationalization indefinitely without making their own contribution. The greatest risks, of course, are posed by the largest organizations, because a strike in them can affect the total economy. In recent years many unionized employees have been paid not so much for the work that they do as for the disruption that they can cause by withdrawing their services.[36]

A case in point is provided by a comparison between Canada Post and the U.S. Postal Service, which is clearly to the credit of the latter. Unlike their Canadian counterparts, who have taken full advantage of their ability to hold the Canadian public hostage, employees in all U.S. federal services are prohibited from capitalizing on strikes. In a report of July 13, 1987, the following comparisons were made using Canadian dollars. U.S. letter carriers received $19.51 against a rate for Canadians of $13.41 per hour. The American postal service charged twenty-nine cents against the Canadian charge of thirty-six cents. Canada Post has been accumulating colossal deficits, whereas the U.S. Postal Service has shown a profit. Canada Post, when it is not on strike, generally takes almost two weeks to deliver a letter from New York to the western provinces. The report concludes: "But more fundamental, and therefore more important, is the poisonous adversarial attitude which has developed over recent years between our post office and the unions representing the postal workers."[37]

Basil Henriques, a much-admired judge in London's East End, used to argue convincingly that one could forecast a young person's future from his or her truancy record at school. Much the same might now be forecast from the rate of industrial absenteeism in the leading Western countries of the Northern Hemisphere, including Japan, with which we have thus far been mainly concerned. In 1987 the absenteeism rate was 2.5 percent in Japan; 3 percent in West Germany and Sweden; 4.7 percent in the United States; 5.5 percent in Holland; 5.9 percent in France; and 7.7 percent in Denmark. But in Great Britain the rate was an alarming 11.8 percent; and in Canada, which has acquired many of its trade-union leaders from Britain in recent years, the rate was 11.6 percent. These figures require no further comment. In Britain and in North America, rationalization has thus far allowed the common man and woman to live in the best of all possible times. If we are to try to retain our present standards, we should perhaps learn to be just a little more flexible and conciliatory among ourselves.

XI Human Equality

It is customary to complain about the continuing inequalities in the material well-being of our species. Certainly such inequalities do exist, but our present acute awareness of their need to be rectified is an indication of how far we have come. Until quite recently, these inequalities remained sanctified by church and state. Although the lines were quietly dropped during my lifetime, the Christian hymn *All Things Bright and Beautiful* still contained the following stanza in *The Book of Common Prayer* published as late as 1908:

> The rich man in his castle,
>   The poor man at his gate,
> GOD made them, high or lowly,
>   And order'd their estate.

As we have seen with time, calendar, chronology, numbers, language, and production, we really do not appreciate to what degree rationalization has already taken place in our human environment until we look back at conditions in the past. Under the feudal system our social structure took the form of a pyramid. This involved having very few people near the top, with large numbers of free men and serfs toward the base. That social structure conformed closely with the Great Chain of Being, in which God and his angels were ranged in order above, and all mortal beings below had their appropriate places. Though the Great Chain of Being comes down to us from Plato and Aristotle and was first systematized by the Neoplatonist Plotinus (ca.

205–70 A.D.), its greatest influence on Western thought was experienced in the Renaissance and the seventeenth and eighteenth centuries. This is understandable. During the latter period, those who were well placed on the Chain of Being had good reason to fear that they might lose some of their privileges through the increasing equality that ultimately results from the urbanization of human beings. We should not be surprised to find that the Great Chain of Being was being most strongly advanced at the very time when man himself was beginning to be emancipated.

As late as 1732, in Epistle 1 of his *Essay on Man*, Alexander Pope exalts the conservative values implicit in the Great Chain:

> Vast chain of being, which from God began,
> Natures aethereal, human, angel, man,
> Beast, bird, fish, insect! what no eye can see,
> No glass can reach! from Infinite to thee,
> . . . .
> Where, one step broken, the great scale's destroy'd.

What is more, within each category, such as within the category of human beings, everything must rise "in due degree."[1] In the middle of the eighteenth century, however, Voltaire and Samuel Johnson spoke out against this concept. And when the Great Chain of Being could no longer be maintained as a viable principle, a new criterion emerged, which was more acceptable to the Sir Richard Arkwrights of this world—namely, the survival of the fittest. This concept, first proposed by Malthus, led directly to Darwinism and much that we have already discussed.

THE RATIONALIZATION OF HUMAN BEINGS

The first cracks in the feudal system had come much earlier than the seventeenth century, and for reasons other than the development of technology. The bubonic plague first reached Sicily in 1347, through Genoese ships arriving from the Crimea. Neither prayer nor the mass burnings of Jews, who were held to have spread the disease by poisoning the wells, could hold it back. It is estimated that in some parts of Europe as many as two-thirds to three-quarters of the population died. The chronicler Jean Froissart (ca. 1337–1410) estimated the deaths at about one-third of the world's population. J. F. K. Hecker has since calculated that twenty-five million people, one-quarter of the population of Europe, died in the epidemic.

In England the Black Death marks a distinct but erratic movement in the direction of greater equality. The immediate effect was a reduction in rents and hence ruin for many landowning families. For the lower orders, however, there resulted an increase in wages and a migration of many peasants into the towns, despite the active disapproval of their betters. E. T. Donaldson notes that Chaucer, in the latter part of the fourteenth century, specifically praises the Parson and his brother, the Plowman, because—like William Langland's *Piers Plowman*—they remain in the countryside to carry out their duties: "The depopulation caused by the great pestilences that made parish priests abandon their duties also made farm workers abandon theirs. Agricultural labourers were legally bound to the soil in Chaucer's time."[2]

The five guildsmen in Chaucer's Prologue to *The Canterbury Tales*—the haberdasher, carpenter, weaver, dyer, and tapestry maker—indicate the dissolution of social stratification that can occur when people move into the city. Chaucer's guildsmen wear expensive new clothes, and they have silver mounting on their knives, when only brass is permitted for their class. This demonstrates to Chaucer that the newly rich guildsmen are anxious to sit on a dais in a guildhall, and that their wives are more than ready to be called "madame."[3] The men who are undercut by Chaucer belong to the guilds whose stranglehold on trade will be broken in the Restoration and eighteenth century. At that time there would be another great step toward fluidity and equalization in the relationships between human beings. It would be a great mistake, however, to imagine that the pestilence of the fourteenth century was, of itself, enough to give people that right to move freely in search of employment, which is the essential prerequisite to equality among people and nations.

It is generally considered that it took until the end of the sixteenth century for the population of Europe to return to its size in the middle of the fourteenth century, before the Black Death. R. R. Palmer estimates that France had twenty-one million inhabitants in 1700, though there had been twenty-three million in the first part of the fourteenth century. In 1700 France was four times as populous as England and twice as populous as Spain.

Until the death of Louis XIV in 1715, only the "unprivileged" classes paid direct taxes in France. This of itself demonstrates the equalization of human beings that has taken place since then. Today, in a complete reversal of past practice, most Western income tax is what is euphemistically called "progressive." This means that rates of personal income tax increase in a proportion *greater* than the increase in wealth. Although he failed to break down internal tariffs completely, Louis XIV's great minister Jean Baptiste Colbert (1619–83) managed to eliminate internal tariffs in the area of "the Five Great [Tax] Farms," in central France. This produced the largest free-trade area in Europe— about the size of England.[4]

In general, however, Europe's internal free trade lagged well behind England's, which had gained greatly from the internal rationalization of its trade. England's human franchise was also ahead of continental Europe's, in the sense that its farmworkers were by no means landless peasants. Of England's population of perhaps five million in 1700, some two million may have belonged to the class of wage-paying tenant farmers. They, in their turn, supported some two and a half million of the class of farm laborers. But life for the ordinary laborer was by no means as golden as the nostalgic brush has been inclined to paint it. Certainly, the wars of the Interregnum had confirmed the centralizing power in London, thereby undermining for a time the ambitions of provincial cities and landowners. But this did not benefit mere peasants. In fact, as James Burke has pointed out, the Acts of Settlement of the mid-seventeenth century had made the village a virtual prison. Unless the justice of the peace—who, as we know from Fielding's novels, was generally the local squire—issued peasants a certificate of movement, they remained confined to the village.[5]

We have earlier noted how the Great Plague of 1665 and the Great Fire of 1666 created an unprecedented demand for goods. The dislocation that ensued from the Great Fire was ultimately of benefit to the rationalization of British manufactures. This dislocation came at approximately the same time as the demystifying of the mysteries or trades, which resulted from the precepts of the Royal Society, founded in 1662. As a result, we find trades beginning to break away from the restrictions of the guilds. This took place, first, when tradesmen went to areas like Clerkenwell outside the City limits. From about 1760, however, the industrial base moved to the North of England, and this marks the beginnings of the British industrial revolution.

The human fodder for the industrial revolution was provided by laborers and tenant farmers driven into the cities by the enclosure of the common land, and by the new, less labor-intensive agricultural economy. Although at first this meant that the lot of the working people as a whole became even worse, they were now only enslaved by their desperate need for wages. The fittest could do more than survive; they could even prosper beyond the dreams of any farm laborer. It should also be noted that workers were not liberated from an agrarian fate until their labor was no longer needed. We shall see parallels later when we look at the more recent enfranchising of blacks and women in North America. Another parallel is that the enfranchised agricultural workers discovered soon enough that personal freedom brought few material perquisites without the concomitant political franchise.

The battle for political franchise was not easily won. G. M. Trevelyan describes the so-called Peterloo massacre. This occurred when some sixty thousand unarmed men, women, and children gathered to demand universal suffrage in Saint Peter's Fields, Manchester, on August 16, 1819. After the cavalry charged, eleven people were killed and several hundred were wounded.[6] An earlier report by J. L. Hammond and Barbara Hammond says of the Peterloo massacre: "The town they met in, though almost the largest in England, was unrepresented in a Parliament that gave two seats to Old Sarum."[7] Old Sarum—without a population base, as a result of the bishop's see having been removed to the present Salisbury in 1220—provided an extreme example of the way in which political representation had not followed the transfer of population.

Eventually, political pressure from the masses led to the Reform Bill of 1832. Before that, no new boroughs had been created since 1688, despite the massive shift of the population to the new industrial towns. The Reform Bill made 143 new seats available by eliminating or reducing the seats in the older or "rotten" boroughs. The resentment the reform caused among the upper classes is indicated by the fact that the House of Lords refused to pass this very limited franchise. They were only persuaded when the king promised to create enough peers to outvote them if they did not submit.

The ex-Jew, Benjamin Disraeli, who knew something about lack of franchise from personal experience, passed the Reform Bill of 1867. Though there was still a ten-pound minimum annual value of lodgings required for the franchise, 938,000 names were added to the electoral register, almost doubling the previous total. At the passing of this bill on July 15, 1867, Robert Lowe, Viscount Sherbrooke, made a famous speech in the House of Commons. Its message, with reference to the new electors, has been popularized as: "Gentlemen, we must now educate our masters."

The political franchise resulted in other forms of franchise to which we have more recently become accustomed. The Reform Bills of 1832 and 1867 lead directly to the Elementary Education Act of 1870. A number of other acts followed almost immediately, and reflected the second Reform Bill in moving toward the equalizing of opportunities for individuals. In 1871 the University Tests Act threw open the older universities to Dissenters; in 1872 the Ballot Act introduced the secret ballot; and in 1875 the Employers and Workmen Act placed masters and men on an equal footing with regard to breach of contract. There was also new legislation that permitted peaceful picketing.

WESTERN TECHNOLOGY AND THE FRANCHISE OF THE UNDERPRIVILEGED

The franchise makes people equal and therefore, in effect, makes them interchangeable. What we call franchise among people is therefore somewhat analogous to what we call rationalization when we are discussing production, or weights and measures. We are perhaps not too happy to think of ourselves as developing into interchangeable parts in an increasingly rationalized world. For better or for worse, however, that is the fate to which our greed and our actions are condemning us.

So much has changed during the past century that we must remind ourselves that the Reform Bill of 1867 refers only to men and only to England. Christian countries and the institution of slavery remained remarkably compatible until well into the nineteenth century. Slavery was not abolished in the British colonies until 1833, in the French colonies until 1848, and in the United States until the Thirteenth Amendment, in 1865. Serfdom remained in the Hapsburg possessions until 1848 and in Russia until 1861. The Latin American republics abolished slavery during the nineteenth century, though it remained in Cuba until at least 1885 and in Brazil until 1888. Nevertheless, peonage, in one form or another, has continued in South America until our own times.

One cannot exaggerate the importance of slavery and slave-produced raw materials to the British industrial revolution. Yet forced labor has a remarkable facility for reducing the productivity of both the workers and their owners. Though the population and the productivity of both the North and the South of the United States were approximately equal in 1790, cheap immigrant labor was making the North far more dynamic by 1860. However, the cotton gin, patented by Eli Whitney in 1794, abruptly revived the almost decaying plantation economy of the South and made it an adjunct to England's industrial revolution.

As Palmer puts it, "British imports of raw cotton multiplied fivefold in the thirty years following 1790. In value of manufactures, cotton rose from ninth to first place among British industries in the same years. By 1820 it made up almost half of all British exports."[8] The agricultural revolution had allowed England to transform its surplus labor into urban wage slaves. Similarly, the cotton gin ultimately contributed to the freedom of the blacks by allowing for the possibility of a wage-based economy in the South. However, the real history of black emancipation belongs to our own time. As with the industrial workers in England, real suffrage did not come for American blacks until public demonstrations led to political suffrage and improving levels of education.

We are concurrently experiencing similar developments related to the suffrage of women. As with liberal-minded politicians in respect of the Reform Bill, or abolitionists in respect of slavery, one can credit suffragettes with the increasing suffrage of women. But in all these cases the work of reformers is only part of the story. Suffrage, like rationalization, can be helped along, but it tends to come only when the world is ready for it. At the risk of offending the majority of feminists, one might suggest that the electric motor—North American households each have some thirty or forty of them—has done at least as much to emancipate women as Emmeline Pankhurst in England or Alice Paul in the United States. Other technological contributions to feminine freedom came from the bicycle, which allowed women to escape temporarily from the home environment after the late nineteenth century, and from the electric starter, which introduced them to cars after 1912. The most immediate cause of the franchise was, of course, the contribution of women in World War I. Though the franchise came in 1920 in the United States and in 1928 in Britain, real equality in education and in the workplace has only very recently become evident.

We are currently going through a transition in North America—and indeed throughout the Western world—which is a modern version of the melting pot. The Hispanics have entered as cheap labor from the outside and are being integrated much as other immigrants have been. But women, blacks, and the handicapped are now being integrated after having long been disadvantaged in their own country, in much the same way as the Jews were integrated thirty or forty years ago. A new invention, like printing or the motor car, has to prove, at first, that it can look and act remarkably like the manuscript or horse-drawn carriage that it proposes to supersede. Similarly, Jews, blacks, women, and the handicapped, who aspire to be recognized and integrated, must look and act much like the existing workforce. But, in time, as they are rationalized and integrated into the system as a whole, the norms that the system demands will change to reflect the composition of the new mix.

Much of what we are currently hearing about illiterate adults is really a reflection of the urgent need to retrain workers trapped in dying industries and to integrate minorities into the system. When we learn, for example, that in the United States four million black, eight million white, and seven million Hispanic adults are illiterate in English, we are dealing with an urgent need for social rationalization. On a smaller scale, British Columbia, with a population of over two and a half million, has just over 10 percent who are adult illiterates. In order to become interchangeable parts within the social mechanism, people must be prepared to train for such jobs as are available. But training for jobs no longer means the apprenticeship system, in which much can be learned through imitation rather than through reading and writing. Moreover, one can no longer aspire to upward mobility without being both literate and numerate. The changing nature of modern industry—particularly the movement away from heavy industry and toward communication and service industries—means that a young person now entering the workforce must expect to be equipped for making several career changes during the course of a lifetime.

On the whole, the countries of the Western world are currently moving very quickly toward full social integration. Even the women in

Switzerland now have the vote. Japan may, perhaps, prove the last bastion of male domination. Nevertheless, twenty-four million Japanese women represent 40 percent of the work force. In 1986 their earnings were only 52 percent of what men made,[9] yet the overall trend to rationalize labor suggests that this will change. We tend to overlook the remarkable extent to which our technological revolution has already rationalized the wage levels of workers within each nation. In England, when and where this all began, Daniel Defoe complains about maidservants "who were formerly hired at three pounds to four pounds a year wages," now costing "above forty shilling a head *per annum*, more."[10] At much the same time, Defoe's Moll Flanders refers to a cloth dealer's "servant maid at 3 pounds a year wages or thereabouts." In the early eighteenth century the minimum income for a squire was five hundred pounds per annum. This meant that his minimum income was between 50 and 150 times that of employees.

Though we may deny it vehemently, the wage differential has been steadily decreasing since Defoe's time. Today, a person in middle management would hardly expect to be earning more than two and a half times the wage of a union worker. And this is even before the further leveling caused by "progressive" taxation and death duties is taken into account. We complain that 20 percent of the people in North America are underprivileged, but we conveniently forget that 80 percent of the people in Greece and Rome were slaves. Even those on welfare in the Western world are far better off than the average worker in the Third World.

Since Defoe's time we have virtually eliminated cheap labor in the Western world and instead contract out for the use of cheap and productive labor abroad. By such actions, we spread the material advantages of technology beyond our shores and ultimately create an even larger market for material goods. Let us consider what this is doing to our world as a whole. At the time of Christ there were about a quarter of a billion people on earth. As a result of wars and plagues, life was then much more nasty, brutish, and short than it is now, and it took until 1650 for that population to double. By 1798, when Malthus was derided for his pessimistic view of the population explosion, the world's people had doubled again to one billion. Within a hundred years it was two billion, and within little more than a further eighty-five years it was five billion. These are compelling figures. They make the projections even more frightening. By 2045 we may have cities (urban conglomerates) of more than a hundred million people, and certainly we can anticipate a population of between ten and fifteen billion people during the course of the next century. Worse still, these are not people who will be happy to live at the same material level as the maidservants who earned three pounds per year in the time of Defoe.

We all now expect to live at a better level than the squire, and—through transportation, refrigeration, and the use of chemicals—enjoy maggot-free and standardized meat, fruit, and vegetables both in and out of season. Some of us complain bitterly, but there never will be a time when people live at a higher level of material wealth than that which rationalization has now brought to the one billion of us in the Western world. This means that our present global population—once we have spread the franchise and the wealth around—will extract some five hundred times as much out of the earth as it did in 1650 (ten times the population multiplied by fifty times the average use of material goods for each person). That is before the projected expansion of the population in the next century to two or three times its present number. We are already witnessing the population explosion that takes place as poorer nations, with their larger families, begin to enjoy rationalized food production and rationalized medicine without the compensating effect of the widespread birth control that occurs in wealthy nations. Since the turn of the century, life expectancy in the developed world has risen from forty-six to seventy-one for males and from forty-eight to seventy-eight for females. As late as the middle of this century, life expectancy in India was still averaging thirty-four; its present rise is further aggravating the population explosion. The problem with human beings is that they constantly condemn rationalization, but they insist on greedily consuming the material wealth that derives directly from it. And the urgent claims of the Third World to share equally in the material benefits that we derive from rationalization cannot be disregarded for long.

Those of us who enjoy the material benefits of the Western world ignore the extent to which this derives from the many forms of rationalization with which this study has been concerned. We complain, however, about the extent to which we are all being propelled into becoming like one another. J. M. Roberts asks the rhetorical question, "Why do we cherish and savour human variety?" He answers, "Perhaps because [variety] is in danger as never before. . . . A new uniformity of human experience is in the making, apparently universal in scope. The world has recently—perhaps even suddenly—begun to seem much less varied . . . more bland, and perhaps a little more grey."[11] The causes, as this study will have shown us, are the rationalization and uniformity that technology demands. One of the reactions to this uniformity has taken the form of what I have called the "ethnic" revival, the desire to appear different while remaining much the same.

The desire to stress variety and yet retain the advantages of technology can be well illustrated by my experience as a wholesale watch importer during the fifties. The German center of watch manufacturing was in Pforzheim. The industry there produced what appeared to be many thousands of distinctive varieties of watches, each of which differed as to case, dial, bracelet, and brand name. Imagine my surprise when I learned that all the factories used between them only eight standard *ébauches*, which are the movements in the rough before the escapement has been fitted and the train has been pivoted. Forty years later, the quartz movement has rationalized the watch trade even further, without in any way reducing the apparent variety in which watches can be purchased. So it is with us. We feel that reason and free will allow us to exhibit unlimited variety as individuals even while our actions remain motivated and predestined by the relative uniformity that technology demands.

The paradox of apparent variety based on standardized production methods extends not only to our consumer products, like clothing and household goods, but also to the houses and towns in which we live. The introduction of standardized double-hung wooden sash windows from Holland during the late seventeenth century resulted in the widespread adoption of Georgian residential architecture, first in Britain and then in North America. This led to the further standardization involving terraced Georgian homes with identical fronts but differing interiors. Early examples are the Bath Circus, started by John Wood, Sr., and finished by his son, as well as their Royal Crescent, built for the Bath of 1775.

Terraced homes provided one form of town planning, particularly during the eighteenth and nineteenth centuries, but another form of standardization had particularly widespread implications for North

America. This was the grid system for streets, which was influenced by Edinburgh's New Town, built between 1768 and 1850. The grid system, which permeates North American town planning, permits a considerable variety of residential architecture within a standardized framework. But even this apparent variety incorporates room heights, stud work, electric circuits, hardware, turnery, and much else which are far more uniform than they appear to be. The Western world, through widespread advertising, has been able to stimulate and channel the demands of its citizens for fashionable (read "relatively uniform") products. This has facilitated, with admirable success, the application of reason and rationalization to their production methods.

Human beings have long claimed, with Descartes, that reason sets them apart from the rest of nature.[12] They are not, however, prepared to make the rational decisions that might allow them to retain the felicitous material condition to which they have been brought by rationalization. A whole series of studies since that by the "Club of Rome" have described the problem.[13] But one hardly needs studies to recognize that, even without the nuclear Armageddon, there lies ahead the inevitability of chemical pollution, carcinogens, infected water, dangerously accumulating waste, and much else if we do not change our present pattern. Our chlorofluorocarbons are already attacking the ozone layer, and the greenhouse effect from our greatly increased emissions of carbon dioxide may well raise our average temperature as much in the next sixty years as in the whole period since the last ice age.

Those who are not intentionally blind know that the problems are already intensifying. Does this mean that we should re-establish the period before rationalization, when the few had much and the many had very little? That is certainly one way of reducing the pressure to create Rachel Carson's world of the *Silent Spring*, but it is not the only way. We who are the heirs of the Judaic tradition were once taught that we should "be fruitful and multiply and replenish the earth, and subdue it: and have dominion over the fish of the sea, and over the fowl of the air, and over every living thing that moveth upon the earth" (Genesis 1:28). That human hegemony has long since been achieved, but we refuse to recognize that the earth can no longer support our species in the manner to which more and more of us are fast becoming accustomed. Furthermore, we seem unable to apply reason to the reduction of our numbers, or our material demands, or both. Seemingly, a vociferous minority, still hidebound by outdated injunctions, refuses to recognize the need for limiting human population, and the rest of the world is unable to apply reason to its own pressing predicament. Is it really true that those whom the gods are about to destroy they first make mad?

Conclusion

Our main concern in this study has been with the wide-ranging rationalization that has taken place during the past three hundred years. We began by looking at the rationalization of temporal measurements. Before the advent of mechanical clocks, hours were measured by dividing either the daylight or the night by twelve. Such an irrational measure of time might do well enough for an agricultural society, but it could never have facilitated the industrial revolution and the spread of Western science and technology.

From the fourteenth century onwards, mechanical clocks brought about the rationalization of the day into twenty-four equal hours. The next major temporal rationalization came through the advent of the pendulum escapement in clocks. This invention, in 1657, resulted in a sixty-fold increase in accuracy; it meant that clocks were, for the first time, sufficiently precise to regulate modern urban life. The related British horological revolution of 1660–1760 ends when, in 1761, John Harrison's fourth chronometer erred by no more than fifteen seconds after a five-month journey to the West Indies and back.

The new increase in horological accuracy called into question the very solar day of the sundials by which clocks had hitherto been regulated. As a result of what is known as the equation of time, it now became evident to the newly time-conscious London public that the solar day varied in length by as much as thirty-two minutes over the course of the year. This brought about the next step in the rationalization of time measurement, which involved the establishment of the mean solar day.

By the early nineteenth century, the railways and the telegraphs were making it all too apparent that the time in London was by no means the same as that in other parts of Britain. This called for the further rationalization of making London time prevail throughout the land. But Britain was only a small country, and railways were beginning to cut their paths across the continents of the globe. A global standard for time measurement had become essential. The need resulted in a further rationalization, which brought us standard time and its twenty-four international time zones that we have today. Standard time, together with the related use of the Greenwich meridian, has been one of the most important milestones on the road toward our global village.

The earlier increases in the accuracy of time measurement were concerned with the search for the longitude and related improvements in marine navigation. Our current rationalization of time measurement is similarly motivated, but navigation is now concerned with the solar system rather than being confined to the earth's surface. Since 1967 the atomic second has been defined by the vibration of cesium 133 atoms, and hydrogen masers now measure time to within infinitesimal parts of such a second.

The Western consciousness of time, which led to the industrial revolution, also led to attempts at rationalizing both the calendar and chronology. The Julian and the subsequent Augustan rationalizations of the earlier lunar calendar did well enough in terms of aligning themselves with the solar year. They failed quite miserably, however, when it came to instituting a rational subdivision for the year itself. Some earlier calendars had been more successful in this respect. The Egyptians, for example, had divided the year into twelve months of thirty days each, followed by five blank days. All months could therefore contain three weeks of ten days each. However, the attempt to align the solar calendar with lunar months (as with the movable feast of Easter), or with the seven-day weeks of Jewish, Christian, and Muslim Sabbatarians has made any subsequent rationalization of the divisions of the calendar all but impossible. Though there has been a genocide of saints' days in Europe, as the work ethic of the Protestant North has taken over from the Catholic South, even the French and the

CONCLUSION   233

Russian revolutions have failed in their assault upon the seven-day week.

The Catholics certainly scored a temporal victory over the Protestants when the Gregorian calendar of 1582 was ultimately accepted by the British in 1752. But implicit in the Gregorian calendar were the seeds of a temporal rationalization that would embarrass Catholics and Protestants alike. When the pope's secretary asked him to look into calendar reform in 1514, Copernicus knew that he must first understand the relationship between the motions of the sun and the moon. From this investigation there ultimately came his *Revolution of the Heavenly Spheres* (1543), and a most important revolution it entailed. People like Galileo began to question the whole Aristotelian and Ptolemaic cosmology, which the Roman Catholic church had made sacrosanct. Worse still, the increased consciousness of time and chronology in the seventeenth century was making even churchmen interpret the temporal implications of the Bible quite literally.

Consequent to Bishop James Ussher's *Annales Veteris et Novi Testamenti* (1650–54), it was now possible to date one's chronology from the precise moment at 9:00 A.M. on Sunday, October 23, 4004 B.C., when the creation occurred. Not only was this date actively disseminated by Christians, but it complemented the increasing rationalization of chronologies by measuring them from the benchmark epoch of the incarnation of Christ. From the seventeenth century onward, this new and increasing consciousness of time was not restricted to matters ecclesiastical. Before long, it was impossible to avoid a serious questioning of whether the creation really could have occurred only six thousand years ago.

Since the British horological revolution, the consciousness of a progress through time has permeated the disciplines and philosophies of the world. We find it in the laissez-faire philosophy of Adam Smith; in the geological stages of Buffon; in the progressive human epochs of Condorcet; in the stone, copper, and iron ages of Thomsen; in the anthropology of Morgan; in the progressive strata of William Smith; and in the inorganic physical processes of Lyell. Also, in the nineteenth century, the sense of progress through time is the basis for the Hegelian and Marxian dialectics, the Darwinian survival of the fittest, Schliemann's digging in Troy, and Freud's descent into the mind. If these expansions in our sense of chronology were not enough, the Einsteinian expanding universe of our own century has made it seem almost unchivalrous to remind ourselves of Ussher's six thousand years. But in many ways our evolution from Ussher's chronologies has served us well. The rationalization of chronology has given us a very real sense of parallels in time through which to compare concurrent developments in the evolution of our universe, our earth, and our civilizations.

In the past our irrational measures of time, calendar, and chronology placed serious limitations on any form of experimental science. But our measures of length, area, weight, and volume were every bit as irrational. Time and the calendar were at least controlled by observable solar phenomena. Man himself, however, with all his variations in size, provided the first benchmark for the measure of length, and the terms used for such measures still reflect their human origins.

During the eighteenth and nineteenth centuries, the British, as the world's leading traders, made some very real attempts to standardize their weights and measures, but they proceeded by evolution rather than by revolution. Toward the end of the eighteenth century, the French, whose weights and measures were even more chaotic than those of the British, took a revolutionary approach. Their metric system was entirely divorced from the measure of man. It was based instead on the meter, which was ostensibly derived from a ten-millionth part of the quadrant of the earth's meridian. Theoretically, though not in practice, their measures of length, area, weight, and volume were all reproducible from the size of the earth and the related meter.

As a result of our modern craving for exact measures, virtually all nations except the United States have now accepted the metric system in principle. Another result is that the meter (just like the second) has been divorced from its previous relationship with the earth. In 1960 the meter was redefined in terms of the wavelength of radiation from a transition in atoms of the element krypton. Even the inch has been similarly redefined as equal to 41,929.399 wavelengths of krypton 86, measured at 760 millimeters pressure and 15 degrees centigrade. More recently, the meter has been defined as the length traveled by light in a specified time. The present definition of the inch in metric terms strongly suggests that the hegemony of the rationalization involved in the metric system is now virtually inevitable.

CONCLUSION

The rationalization of numbers is a prerequisite for the rationalization of all the measures with which we have been concerned. Clearly the decimal-based value of the Hindu-Arabic numerals provides a natural adjunct to the metric system. But it was not always this way. Certainly the ten-part numbering system is related to primitive finger counting, just as the twenty-part numbering system, by scores, includes the toes. But there is also a quite different system of duodecimal- and sexagesimal-based numbers. The duodecimal system is still very much with us in the months of the year and the hours of the day and the night, as well as in counting by the dozen and the gross. The sexagesimal system still remains in the circle of 360 degrees, and in the minutes and seconds into which both degrees and hours are divided.

There is nothing intrinsically inferior about the duodecimal or sexagesimal number systems. Their quietus has probably been hastened by the importation into Europe of the Hindu-Arabic numerals, via the Muslims of Spain, at the beginning of our own millennium. The subsequent addition of the zero and the decimal point has made this system an integral part of the language of Western science, technology, and commerce. This rationalized number system, which we call the Hindu-Arabic numerals, has now become the unquestioned *lingua franca* of the modern world. Moreover, the integration of the number system into the new temporal-based mathematics, such as dynamics and the differential calculus, has made it an essential tool of the modern time-based disciplines.

Today, few people question the rationalization that has given us our temporal measurements, our metric system, and our Hindu-Arabic numerals. There is, however, still some hesitation on nationalistic grounds about the acceptance of English as the international medium of communication. The beginnings of English would hardly seem to have been propitious for the founding of a world language. Perhaps the most important quality of English is that it has always been a scavenger language built up on the laissez-faire lines that the British were later to prize in their trading practices. The original language of the Germanic invaders gave English the simple words of everyday speech in an agricultural economy. The entry of Christianity, in the sixth and seventh centuries A.D., brought an infusion of Latin and the more sophisticated ideas that its words could convey. When the Danes lived alongside the Saxons, after the middle of the eighth century, a pidginization of the two relatively similar languages took place. As a result, Old English's heavily inflected forms became very much simplified.

The Norman invasion of 1066 enriched Middle English by no fewer than ten to twelve thousand words. During the English Renaissance a similar number of imported words was added from sources as diverse as the Greek terms used by early scientists and the words imported by Elizabethan seamen. During the latter half of the seventeenth century the language was further enriched by a whole host of technical terms. Encouraged by Francis Bacon and the Royal Society, the indigenous "mysteries" were written down, and became the trades that all could learn. There was a concurrent pressure from the early scientists for a "mathematical plainness" and an easy, lucid style in the language. This—together with the rationalization of spelling, meaning, punctuation, capitalization, and syntax—has made the language of the Restoration and eighteenth century recognizable as the modern English that we use today. The melting-pot nature of England's English spread across the world in the wake of the British Empire. Later, England's English was supplemented by the even more active and widespread scavenging of that other melting pot, the United States.

Today, English is the first or second language of more than 20 percent of the world's five billion people. In international science, commerce, technology, transport, finance, and politics, it has no effective competitor. While France's rationalization of weights and measures will almost certainly make its metric system prevail in our impending global village, in language the French have not similarly succeeded.

The standardization of time measurement, weights and measures, numbers, and language are all essential prerequisites for the standardization of production that we call the industrial revolution. As we follow these developments through, we find that they are all involved in an ongoing symbiosis. The demystifying of the "mysteries" in the late seventeenth century not only enriched the language by adding technical terms, but it also weakened the guilds and thereby opened the door to a rationalization of the trades. The concurrent development of financial institutions prepared the economic environment for the projectors and entrepreneurs who would undertake that rationalization. Furthermore, the consciousness of time, which developed in the same period, permitted that great projector Benjamin Franklin (1706–90) to produce his totally topical equation, "time is money."

CONCLUSION 237

Time and money epitomize the values held by the new industrialists of the eighteenth century. In that century men like Crowley, Wedgwood, Arkwright, Boulton, and Watt laid the foundations for the time and method study through which factory production would be rationalized. Understandably, such proprietors did not tend to advertise the new "mysteries" through which they increased their industrial efficiency. That would not occur until the advent of managers and related professionals, during the latter part of the nineteenth century.

The very watches that epitomize the control of production by time also provide, through their process of manufacture, the best example for the rationalization of production. Before he died in 1713, Thomas Tompion, the father of English clockmaking, had instituted batch production for producing the six thousand watches that represented his life's work. In the middle of the century, Diderot's *Encyclopédie* already describes clearly and in detail twenty-one processes for the production of a watch. When Babbage wrote his *Economy of Machinery and Manufactures* (1832), he reported that watchmaking was divided into 102 different trades. Clearly, the simplification that is a prerequisite for mass production was now in place. But the rewards for taking the next step would go to the so-called American System of Manufacture, with its interchangeable parts. This system could use unskilled labor, much of which came from the unindustrialized parts of Europe. By 1822 Eli Terry (probably influenced by Eli Whitney's mass-production methods for muskets) was making six thousand clocks a year at fifteen dollars each. Twenty years later, Chauncey Jerome began exporting brass clocks to England at a dollar and a half each.[1]

The American miracle had begun, and it was helped greatly by the two world wars in our own century. Today, however, we sense that America is entering into an era of decline. The transfer of Western technology to non-Caucasian peoples has taken far less time than the three centuries required for its development. Also, the short-sightedness of British and American management, as well as the recalcitrance of their unionized workers, is playing no small part in the relative decline of their industry.

Once again, watch-trade statistics and the related implications of time consciousness and mass production illustrate graphically the changes that are now under way. The statistics underline the current industrial dominance of Japan. They also indicate that time consciousness is spreading quickly to the Third World, starting with the four "little dragons" of South Korea, Taiwan, Singapore, and Hong Kong. At the turn of this century, Henry Ford—who was the most avid follower of the industrial engineer Frederick Winslow Taylor and his precepts for rationalized production—reluctantly discarded plans for the mass production of cheap American watches. Ford could simply not envisage the necessary sale of 600,000 pieces per year. According to the Citizen Watch Company of Japan, the total worldwide sale of watches in 1986 was just under 530 million pieces. Of these, the Japanese provided 190 million. Nothing could better illustrate the exponential increase that has taken place in rationalized production. But this also illustrates the compelling temporal values that demand more than one watch per year for every ten men, women, and children on the face of this earth.

The rationalization of production methods has also brought about the rationalization of distribution and finance. In the days of the guilds, goods were frequently sold at the very shop where the master worked. When the factories moved to the North of England, retailing developed further as a trade in its own right. By 1900 retailing had begun to be rationalized in two main directions. One direction was represented by Woolworth's first successful store, in 1879, and the founding of England's Marks and Spencer in 1887. The second direction involved multiple shops of grocers, provision merchants, tea dealers, newsagents, tobacconists, and chemists. These had the advantage of a centralized office and centralized buying, together with the ability to develop as clones, patterned after a successful prototype. In the fifties and sixties of this century, the corporate chain stores controlled an increasing amount of retail business. But they now seem to be fighting a rearguard action, in part because of labor difficulties.

During the past two decades, business-format franchising seems to be taking over in more and more areas of the retail and service industries. This form of franchising combines the advantages of centralized know-how, the cloning of a successful format, and the frenetic entrepreneurial activity of the individual owner. The spreading of the political franchise, which is a concomitant of the spreading of Western technology, has resulted in an increase in potential entrepreneurs, particularly among women and minorities. At the same time, the rising status of many people has created the demand for a whole spectrum of

CONCLUSION 239

service industries. They now provide a rationalized substitute for the work that was previously performed by wives and servants. Three hundred years ago, virtually all of us would have been peasants. Today, the one billion of us who enjoy the fruits of Western technology all expect the same material standards as were previously the privilege of the squire. And Western technology has a happy weakness. The mass production of material goods cannot continue without a growing core of customers; Ford was right to recognize that his workers must also be his clients.

Ultimately, the wider distribution of money means the wider distribution of franchise, and both money and the franchise have been distributed on an unprecedented scale during our own century. First came equal opportunity for men, and, more recently, there came equal opportunity for the women and minorities who had previously been the hewers of wood and drawers of water in the Western world. With equality has come the need to dress, to act, and to speak like the rest of the society into which one aspires to be integrated, to become, in fact, one of the interchangeable parts in the social mechanism of the rationalized Western world. Human beings naturally react against paying such a price, but their desire for the fruits of rationalization is making them become more and more standardized before our very eyes.

In the past there have been reactions—comparable to those opposing the various production methods—against virtually all the forms of standardization with which we have been concerned. The reaction has sometimes inhibited the possibility of a full rationalization, but a compromise has always been achieved in the interest of furthering the material progress of human beings. In time measurement, for example, the traditional duodecimal and sexagesimal systems have been retained in the numbering of hours, minutes, and seconds, for which there is an established precedent. Our more recent cosmic measurements, however, are now made in light kilometers, and for measures below the second we use a series from nanoseconds down to chronons, which derives from the decimal system. In the calendar the seven-day week supported by Sabbatarian reactionaries has prevailed. Such weeks are patently inconsistent with the irregular durations of our months, for which there is equally little excuse. We have, however, now further rationalized the Gregorian system so that it will not differ from the calendar of the sun by more than one day in twenty thousand years.

In chronology, the whole world now shares the useful fiction that Christ was incarnated on the Christmas falling between B.C. and A.D. The chronology of the universe has, however, been extended far beyond Ussher's 4004 B.C.

In weights and measures, much of the reaction against accepting the rationalization of the metric system belongs to our recent memory. In numbers, though we may still play with the nostalgia of dozens and scores, it is clear that the decimal system is fast taking over. Just as Hindu-Arabic numerals now represent the world's *lingua franca*, so English is already used by more than twenty percent of its population. Though the French will fight a long and hard rearguard action, those who wish to be involved, on a global scale, in science, commerce, transport, finance, and politics would do well to gain or retain a knowledge of Standard English. The rationalizing of all these elements—time, weights and measures, numbers, and language—has been an essential prerequisite of the Western technology that has produced our current bonanza in material goods.

We experience the reaction to Western technology every day. Human beings have also reacted very strongly against the mechanization and standardization of themselves that Western technology requires. But their greed for material goods has made them, however unwillingly, come to terms with the related standardization. The recent phenomenon of the hippies of the sixties, who became the yuppies of the eighties, illustrates this dilemma. But the hippies were the children of the Western bourgeoisie, and they reacted, however temporarily, against the material wealth by which they were surrounded and supported.

In our own time, Western technology is spreading beyond its traditional centers. It is moving out to establish itself in the many parts of the world where hard work for low wages is welcomed by young and old alike. The catalysts for this spread of wealth—to be followed by franchise and education—are the international corporations. In their search to substitute cheap and willing labor for the privileged unionized labor of the West, they will ultimately create a new middle class in one Third World country after another. Though we barely recognize it as yet, a concomitant of spreading Western technology is a remarkable leveling of both wealth and franchise among all the people of the world.

CONCLUSION   241

In the West the spreading of wealth and franchise is well under way. At the beginning of the period with which we are concerned, Defoe's servant girls earned three pounds per year, while the minimum income for a squire was five hundred pounds. At the very least, we are dealing here with a fifty-fold differential which has, since that time, been dropping steadily. Today, the difference in income between a unionized worker and a member of middle management—both now privileged classes—rarely exceeds two- or three-fold. This is even before allowing for the further leveling provided by "progressive" income tax. At the same time, the population has increased more than ten-fold from the five hundred million who lived on earth in 1650. This means that when the earth's inhabitants have all achieved the material wealth of a squire (which is to say the average wealth of those who now live in the West), we will extract some five hundred times as much material out of the world as we did in 1650. And that takes no account of the projected two- to three-fold increase in the earth's population during the course of the next century.

Swift's image of satire, as a glass in which we see everybody but ourselves, may help us to understand our current predicament. Why are we so vociferous in condemning rationalization, while insisting that we greedily consume the material wealth that derives directly from it? And why are we condemning rationalization when it is the source of the spreading franchise from which we all benefit? What we perilously ignore are the imminent consequences of a five-hundred-fold increase in what we extract annually from the earth. Those consequences are already reflected in chemical pollution, carcinogens, infected water, and much else. People frequently recognize the advantages of holistic medicine for themselves. But they rarely understand how much more important it is to develop a holistic approach toward the health of their macrocosm, the biosphere that is our world. For three hundred years we have been heavily indoctrinated with the propriety of pursuing material progress through time. To achieve that goal, we have rationalized virtually everything, including ourselves. But we now seem to be quite incapable of rationally limiting the extraction of material goods from the earth, or of limiting the size of the world's population which demands these goods.

Old urges and traditions die hard. But if we cannot restrain them, there will be no need for a nuclear Armageddon before we hand over this world to the bacteria, or to the cockroaches and the ants. In his *Discourse on Method* (1637), and his subsequent letters, Descartes argues that human beings are superior to the animal world. He feels that this is because animals cannot communicate their thoughts through language, and because they lack reason: "If they thought as we do, they would have an immortal soul like us."[2] In the succeeding three and a half centuries, we have progressively rationalized and standardized both our activities and ourselves. We are now becoming much like the clockwork dogs, devoid of human reason and soul, which are central to the Cartesian mechanistic philosophy. Also, our goal is now to enjoy more and more material wealth on this earth without worrying too much about our heirs or our souls. Before the French Revolution, the aristocrats comforted themselves with the words, "après nous le déluge." On a much larger scale, our own actions reflect the same rejection of responsibility more faithfully than we dare admit, even to ourselves.

In our day, old Father Time is, once again, in the process of drawing his daughter Truth out of her dark well. We must all now face her blinding nakedness. For three hundred years we have succeeded in rationalizing, in a seemingly unending progress, our creation of a Faustian cornucopia of material goods. But at last the moment of truth is upon us. We shall discover, soon enough, whether human beings really can exercise the reason through which Descartes differentiates them from the animal world.

Notes

I. TIME AND CLOCKS: THE DIVISIONS OF THE DAY

1. Martin P. Nilsson, *Primitive Time Reckoning* (Lund, Sweden: C. W. K. Gleerup, 1920), p. 42.
2. E. J. Bickerman, *Chronology of the Ancient World* (London: Thames and Hudson, 1980), p. 14.
3. W. M. O'Neil, *Time and the Calendars* (Sydney, Australia: Sydney University Press, 1975), p. 4; and F. A. B. Ward, *Time Measurement* (London: Science Museum, 1970), p. 35.
4. Ward, *Time Measurement*, p. 35.
5. Samuel L. Macey, *Patriarchs of Time: Dualism in Saturn-Cronus, Father Time, the Watchmaker God, and Father Christmas* (Athens: University of Georgia Press, 1987), pp. 15–16.
6. Bickerman, *Chronology*, p. 14.
7. *Manual of Seamanship, 1937* (London: His Majesty's Stationery Office, 1940), 1:29.
8. Samuel L. Macey, *Clocks and the Cosmos: Time in Western Life and Thought* (Hamden, Conn.: Archon, 1980), pp. 21–25.
9. Bickerman, *Chronology*, p. 16.
10. Ward, *Time Measurement*, pp. 35–36; Macey, *Clocks and the Cosmos*, p. 22; and John Read, "Daimyo Clocks," *Horological Journal* 130 (June 1988): 12–13, 20–21.
11. Bickerman, *Chronology*, pp. 14–15.
12. Eviatar Zerubavel, *Hidden Rhythms: Schedules and Calendars in Social Life* (Chicago: University of Chicago Press, 1981), pp. 35–36.
13. John Trevisa, trans., *On the Properties of Things* (Bartholomoeus

Anglicus De proprietatibus rerum) (Oxford: Oxford University Press, 1975), pp. 529–35.

14. Ibid.

15. Christiaan Huygens of Zulichem, "Horologium" (1658), trans. Ernest L. Edwardes, *Antiquarian Horology* 7 (December 1970): 43–44.

16. Michael Hurst, "The First Twelve Years of the Pendulum Clock," *Antiquarian Horology* 6 (June 1969): 146.

17. D. W. Waters, "Time, Ships and Civilization," *Antiquarian Horology* 4 (June 1963): 85.

18. Eric Bruton, *Clocks and Watches* (Feltham, U.K.: Hamlyn, 1968), p. 84.

19. Samuel L. Macey, "The Early History of Chronometers: A Background Study Related to the Voyages of Cook, Bligh, and Vancouver," *B. C. Studies* 48 (Summer 1978): 14–23.

20. John Stowe, *The Survey of London*, 3d ed. (London, 1618), pp. 792–93, cited from Thomas R. Smith, "Manuscript and Printed Sea Charts in Seventeenth-Century London: The Case of the Thames School," in *The Compleat Plattmaker*, ed. Norman J. W. Thrower (Berkeley and Los Angeles: University of California Press, 1978), p. 50.

21. Derek Howse, *Greenwich Time and the Discovery of the Longitude* (Oxford: Oxford University Press, 1980), p. 82.

22. Ward, *Time Management*, pp. 1–2.

23. Howse, *Greenwich Time*, p. 82; Alan Smith, ed., *The Country Life International Dictionary of Clocks* (New York: Putnam's, 1979), pp. 324–25.

24. Smith, *Dictionary of Clocks*, pp. 38, 268; Cecil Clutton, G. H. Baillie, and C. A. Ilbert, *Britten's Old Clocks and Watches and Their Makers*, 8th ed. (New York: Dutton, 1973), pp. 293, 514. For an equation table issued by the clockmaker Daniel Quare around 1700, see: "For to Keep a Pendulum Watch in Good Order," *Horological Journal* 121 (July 1978): 4–7.

25. John Donne, *The Poems*, ed. Herbert J. C. Grierson (London: Oxford University Press, 1912), 1:275–76.

26. Macey, *Clocks and the Cosmos*, pp. 129, 133.

27. Howse, *Greenwich Time*, pp. 83–84.

28. Cited from Howse, *Greenwich Time*, p. 86.

29. Cited from Howse, *Greenwich Time*, pp. 114–15.

30. William D. Johnstone, *For Good Measure* (New York: Holt, Rinehart and Winston, 1975), pp. 228–33.

31. Marcia Bartusiak, "The Ultimate Timepiece," *Discover*, May 1981, p. 78.

32. Cited from Howse, *Greenwich Time*, pp. 150–51.

33. A. A. O. Fox, ed., *An Anthology of Clocks and Watches* (Swansea, U.K.: Published by the editor, n.d.), pp. 10–11.

34. Cited from Lawrence Wright, *Clockwork Man* (London: Elek, 1968), p. 29.

35. William Blake, *The Poetry and Prose*, ed. David V. Erdman (New York: Doubleday, 1970), p. 214.

36. Charles Lamb, *Essays of Elia* (London: Dent, 1962), pp. 97–98.

37. Samuel L. Macey, *Money and the Novel: Mercenary Motivation in Defoe and His Immediate Successors* (Victoria, B.C.: Sono Nis, 1983), pp. 99–103.

38. Samuel L. Macey, "Shelley and the New Romantics," *Texas Quarterly* 14 (Summer 1971): 91–95.

39. Jonathan Swift, *Works*, ed. Herbert Davis (Oxford: Basil Blackwell, 1965), 11:35.

II. CALENDARS: THE DAYS OF THE YEAR

1. I have previously dealt with part of this subject in Samuel L. Macey, "Cronus in the Eternal City: Time in Rome and England," *Social Sciences* 53 (Summer 1978): 139–46.

2. Samuel L. Macey, *Patriarchs of Time: Dualism in Saturn-Cronus, Father Time, the Watchmaker God, and Father Christmas* (Athens: University of Georgia Press, 1987), p. 22.

3. Martin P. Nilsson, *Primitive Time Reckoning* (Lund, Sweden: C. W. K. Gleerup, 1920), p. 149.

4. The Roman festival of purification was held on February 15; for the Christian parallels see Macey, *Patriarchs of Time*, p. 139.

5. James Burke, *The Day the Universe Changed* (London: British Broadcasting Corporation, 1985), pp. 134–37; I am also indebted throughout this chapter to the excellent article under "Calendar" in *Britannica*.

6. Eviatar Zerubavel, *The Seven-Day Circle: The History and Meaning of the Week* (New York: Free Press, 1985), p. 75.

7. C. W. Ceram [Kurt W. Marek], *Gods, Graves, and Scholars: The Story of Archaeology*, trans. E. B. Garside (London: Victor Gollancz, 1952), pp. 356–58.

8. Georges Ifrah, *From One to Zero: A Universal History of Numbers*, trans. Lowell Blair (New York: Viking Penguin, 1985), p. 411.

9. Ibid., pp. 405–12.

10. Paul Alkon, "Changing the Calendar," *Eighteenth Century Life* 7 (January 1982): 6–7, 12–15.

11. Eviatar Zerubavel, *Hidden Rhythms: Schedules and Calendars in Social Life* (Chicago: University of Chicago Press, 1981), pp. 98–99.

12. Ibid., pp. 83–95; Zerubavel, *The Seven-Day Circle*, pp. 28–29.

13. Zerubavel, *The Seven-Day Circle*, pp. 80–82.

14. Cited from Zerubavel, *Hidden Rhythms*, p. 99.
15. For much of the material on the Soviet calendar, I am indebted to Zerubavel, *The Seven-Day Circle*, pp. 35–43.
16. Iona Opie and Peter Opie, eds., *The Oxford Dictionary of Nursery Rhymes* (Oxford: Clarendon Press, 1952), pp. 380–81.
17. See the chapter on "Saturnalia and Christmas" in Macey, *Patriarchs of Time*, pp. 111–34.

III. CHRONOLOGY: THE YEARS OF THE WORLD

1. Samuel L. Macey, "Literary Images of Progress: The Rise and Fall of a Western Ideal," in *Time, Science, and Society in China and the West: The Study of Time V*, ed. J. T. Fraser, N. Lawrence, and F. C. Haber (Amherst: University of Massachusetts Press, 1986), p. 98.
2. J. T. Fraser, *Of Time, Passion, and Knowledge* (New York: Braziller, 1975), pp. 149, 469.
3. S. G. F. Brandon, *History, Time and Deity* (Manchester: Manchester University Press, 1965), pp. 43–45, 104; R. C. Zaehner, *Hinduism* (London: Oxford University Press, 1966), pp. 61–62; and Daniel J. Boorstin, *The Discoverers* (New York: Random House, 1983), pp. 558–59.
4. Robert Silverberg, *Clocks for the Ages* (New York: Macmillan, 1971), pp. 156–59.
5. Samuel L. Macey, *Patriarchs of Time: Dualism in Saturn-Cronus, Father Time, the Watchmaker God, and Father Christmas* (Athens: University of Georgia Press, 1987), pp. 1–24.
6. E. J. Bickerman, *Chronology of the Ancient World* (London: Thames and Hudson, 1980), pp. 70–75.
7. W. M. O'Neil, *Time and the Calendars* (Sydney, Australia: Sydney University Press, 1975), pp. 10–12.
8. Eviator Zerubavel, *Hidden Rhythms: Schedules and Calendars in Social Life* (Chicago: University of Chicago Press, 1981), p. 97; Boorstin, *The Discoverers*, pp. 596–97.
9. [Thomas Hearne], *Ductor historicus* (London, 1698), p. 35.
10. Cited from Paul Alkon, "Johnson and Chronology," in *Greene Centennial Studies*, ed. Paul J. Korshin and Robert R. Allen (Charlottesville: University Press of Virginia, 1984), p. 164.
11. Samuel L. Macey, *Money and the Novel: Mercenary Motivation in Defoe and His Immediate Successors* (Victoria, B.C.: Sono Nis, 1983), pp. 26–28.
12. [Daniel Defoe], *A Plan of the English Commerce* ([1730]; reprint, New York: Augustus M. Kelley, 1967), pp. 328–34. But see also Daniel Defoe, *A*

New Voyage Round the World, in *Works,* ed. G. H. Maynadier (Boston: David Nickerson, 1903), 14:264–65.

13. [Hearne], *Ductor historicus,* p. 35.

14. Paul Alkon, "Changing the Calendar," *Eighteenth Century Life* 7 (January 1982): 5–12.

15. John Locke, *Some Thoughts Concerning Education* (1693; reprint, Menston, U.K.: Scolar Press, 1970), p. 219.

16. Samuel L. Macey, "Clocks and Chronology in the Eighteenth Century Novel," *Eighteenth Century Life* 7 (January 1982): 96–104 (see in particular n. 15).

17. Macey, *Patriarchs of Time,* pp. 4–10.

18. John Locke, *An Essay Concerning Human Understanding,* ed. John W. Yolton (London: Dent, 1965), 1:157–58.

19. Silverberg, *Clocks for the Ages,* pp. 17–18.

20. Frederick Engels, *Ludwig Feuerbach and the Outcome of the Classical German Philosophy,* ed. C. P. Dutt (New York: International, 1941), p. 44.

21. R. F. Jones, *Ancients and Moderns,* 2d ed. (Berkeley and Los Angeles: University of California Press, 1961), p. 60.

22. Locke, *Human Understanding,* 1:xxxii.

23. Cited from Boorstin, *The Discoverers,* p. 455.

24. Cited from James Burke, *The Day the Universe Changed* (London: British Broadcasting Corporation, 1985), p. 256.

25. William Powell Jones, *The Rhetoric of Science* (Berkeley and Los Angeles: University of California Press, 1966), p. 17.

26. Edward Young, *Conjectures on Original Composition,* ed. Edith J. Morley (Manchester: Manchester University Press, 1918), pp. 19 and 7.

27. Thomas Robert Malthus, *An Essay on the Principle of Population,* ed. Philip Appleman (New York: Norton, 1976), p. 21; see also pp. 48–50 for the primitive demography.

28. Cited from Burke, *Universe,* p. 273.

29. Ibid., pp. 268–69.

30. Silverberg, *Clocks for the Ages,* p. 65.

IV. THE ASCENDANCY OF THE METRIC SYSTEM

1. Bruno Kisch, *Scales and Weights: A Historical Outline* (New Haven, Conn.: Yale University Press, 1965), pp. 4–6.

2. Cited from Kisch, *Scales and Weights,* p. 4.

3. Cited from Arthur Klein, *The World of Measurements* (New York: Simon and Schuster, 1974), p. 29.

4. Hubert Hall and Frieda J. Nicholas, eds., *Select Tracts and Table Books Relating to English Weights and Measures (1100–1742)*, in *Camden Third Series*, vol. 41 (London: Offices of the Society, 1929), pp. 50–53.

5. Ibid.

6. Philip Rush and John O'Keefe, *Weights and Measures* (London: Methuen, 1962), p. 7.

7. William D. Johnstone, *For Good Measure* (New York: Holt, Rinehart and Winston, 1975), p. 8; and Ronald Edward Zupko, *A Dictionary of English Weights and Measures from Anglo-Saxon Times to the Nineteenth Century* (Madison: University of Wisconsin Press, 1968), p. 85.

8. Johnstone, *For Good Measure*, p. 11.

9. Klein, *The World of Measurements*, p. 45 and note.

10. Ronald Edward Zupko, *British Weights and Measures: A History from Antiquity to the Seventeenth Century* (Madison: University of Wisconsin Press, 1977), p. 141.

11. Ibid., p. 171.

12. G. G. Coulton, *The Medieval Village* (Cambridge: Cambridge University Press, 1925), p. 45.

13. Rush and O'Keefe, *Weights and Measures*, p. 60.

14. Cited from Hall and Nicholas, *Select Tracts*, p. 28; see also W. Cunningham, *The Growth of English Industry and Commerce During the Early and Middle Ages*, 5th ed. (Cambridge: Cambridge University Press, 1927), pp. 119–20.

15. Zupko, *British Weights and Measures*, p. 175.

16. A. E. Berriman, *Historical Metrology* (New York: Greenwood Press, 1969), p. 29.

17. Klein, *The World of Measurements*, pp. 55–57.

18. Ibid., p. 84.

19. Mary Dormer Harris, ed., *The Coventry Leet Book; or, Mayor's Register* (London: Kegan Paul, Trench, Trübner, 1907), p. 396.

20. Rush and O'Keefe, *Weights and Measures*, pp. 22–23.

21. Ibid., pp. 24–25.

22. Ibid., pp. 70–71.

23. John L. Feirer, *SI Metric Handbook* (New York: Charles Scribner's Sons, 1977), pp. 1–4.

24. British government report of 1869, cited from Norman Clarke, in *Metrication*, rev. ed., ed. F. W. Kellaway (Harmondsworth, U.K.: Penguin, 1970), p. 12.

25. Feirer, *SI Metric Handbook*, pp. 2–4.

26. *Metric Manual* (Neenah, Wis.: J. J. Keller and Associates, 1974), pp. 23–24.

27. Ibid., pp. 60–65.

28. John Quincy Adams, *Selected Writings*, ed. Adrienne Koch and William Peden (New York: Knopf, 1946), p. 311.

29. Diogenes Laërtius, *The Lives and Opinions of Eminent Philosophers*, trans. C. D. Yonge (London: Henry G. Bohn, 1853), p. 397.

30. Plato, *Theaetetus*, trans. Henry Cary, in *Works* (London: George Bell, 1896), 1:413.

31. Plato, *Cratylus*, trans. George Burges, in *Works* (London: George Bell, 1901), 3:287.

32. James Boswell, *Life of Johnson* (London: Dent, 1949), 2:123.

33. Samuel Johnson, *Preface to Shakespeare*, in *Johnson on Shakespeare*, ed. Arthur Sherbo, in *Works of Samuel Johnson* (New Haven, Conn.: Yale University Press, 1968), 7:78.

34. Boswell, *Life of Johnson*, 2:123.

35. Horace, *Epistles*, trans. James Lonsdale and Samuel Lee (London: Macmillan, 1873), p. 174 (1.7); Charles E. Passage translates the same line: "Each man must gauge things himself by his own proper standard and measure." Horace, *Epistles*, in *Complete Works*, trans. Charles E. Passage (New York: Frederick Ungar, 1983), p. 274.

36. *Metric Manual*, p. 65; Johnstone, *For Good Measure*, p. 7.

37. Martin P. Nilsson, *Primitive Time Reckoning* (Lund, Sweden: C. W. K. Gleerup, 1920), p. 42.

V. THE ASCENDANCY OF HINDU-ARABIC NUMERALS

1. Karl Menninger, *Number Words and Number Symbols: A Cultural History of Numbers*, trans. Paul Boneer (Boston: MIT Press, 1969), pp. 11, 30.

2. "Lumps of Clay That Gave Birth to Numbers," *Discover*, March 1987, pp. 7–8.

3. Ibid.

4. Georges Ifrah, *From One to Zero: A Universal History of Numbers*, trans. Lowell Blair (New York: Viking Penguin, 1985), p. 88.

5. Menninger, *Number Words*, pp. 11–12, 30, 33–34.

6. Daniel Defoe, *Robinson Crusoe* (New York: Signet, 1961), p. 67.

7. Ifrah, *From One to Zero*, p. 17.

8. Ibid., p. 67.

9. Menninger, *Number Words*, p. 201.

10. Ibid., pp. 201–8.

11. Ifrah, *From One to Zero*, pp. 55–57.

12. Publius Ovidius Naso, *Fastorum libri sex.*, ed. and trans. Sir James George Frazer (1929; reprint, Hildesheim: Georg Olms, 1973), pp. 121, 123.

13. Ifrah, *From One to Zero*, p. 64.
14. Ibid., pp. 142–43; Menninger, *Number Words*, pp. 242–43.
15. Menninger, *Number Words*, p. 122.
16. Ifrah, *From One to Zero*, pp. 70–71.
17. J. M. Christoforakis, *Knossos: Visitor's Guide*, 2d ed. (Heraklion, Crete: Athens Publishing Center, n.d.), p. 72.
18. Cited from Menninger, *Number Words*, p. 136.
19. Menninger, *Number Words*, p. 136.
20. Ibid., pp. 418–20.
21. Ifrah, *From One to Zero*, pp. 475–77.
22. Menninger, *Number Words*, pp. 400–402.
23. Ibid., p. 51.
24. Ibid., p. 157.
25. Ifrah, *From One to Zero*, pp. 65–66.
26. Ibid., p. 65.
27. Ibid., pp. 151–53, 168–69.
28. Ibid.
29. J. Bronowski, *The Ascent of Man* (Boston: Little, Brown, 1973), pp. 179–87.
30. Ibid., p. 179.
31. Ibid., p. 181.
32. Ibid., p. 184.
33. James Burke, *The Day the Universe Changed* (London: British Broadcasting Corporation, 1985), p. 151.
34. Bronowski, *The Ascent of Man*, p. 184.
35. Geza Szamosi, "The Origin of Time," *The Sciences*, September–October 1986, pp. 33–39.
36. Burke, *Universe*, pp. 144–45; Szamosi, "The Origin of Time," p. 33.
37. Richard Dawkins, *The Blind Watchmaker* (Harlow, Essex: Longman Scientific and Technical, 1986), pp. 35, 23.
38. Ibid., pp. 95–97.
39. R. V. Jones, "Time and Distance," *Horological Journal* 127 (December 1985): 6–8, 14; and 128 (February 1986): 10–11.
40. Ibid.
41. James Burke, *Connections* (Boston: Little, Brown, 1978), p. 249.
42. Jones, "Time and Distance," p. 14.
43. Marcia Bartusiak, "The Ultimate Timepiece," *Discover*, May 1981, p. 79.
44. Ibid., pp. 79, 82–83.
45. Cited from Hans Reichenbach, *The Philosophy of Space and Time*, trans. Maria Reichenbach and John Freund (New York: Dover, 1958), p. 160.

46. Samuel Alexander, *Space, Time, and Deity: The Gifford Lectures at Glasgow, 1916–1918* (London: Macmillan, 1920), 1:58.

47. Ibid., 1:59 and note.

48. J. T. Fraser, "The Problems of Exporting Faust," in *Time, Science, and Society in China and the West: The Study of Time V*, ed. J. T. Fraser, N. Lawrence, and F. C. Haber (Amherst: University of Massachusetts Press, 1986), p. 6. For a description of superspace, see J. T. Fraser, *The Genesis and Evolution of Time* (Amherst: University of Massachusetts Press, 1982), pp. 170–73.

49. Bronowski, *The Ascent of Man*, pp. 247–52. Stephen W. Hawking, *A Brief History of Time: From the Big Bang to Black Holes* (Toronto: Bantam, 1988), p. 143; for a lucid outline of the history of developments related to cosmology and time measurement see pp. 19–42.

50. Dawkins, *The Blind Watchmaker*, p. 293.

51. Bartusiak, "The Ultimate Timepiece," p. 83.

52. Bertolt Brecht, *The Life of Galileo*, trans. Desmond I. Vesey (London: Methuen, 1960), p. 65 (scene 6).

53. Tony Rothman, "A 'What You See Is What You Beget' Theory," *Discover*, May 1987, pp. 90–99. See also Hawking, *History of Time*, pp. 124–26.

54. Samuel Taylor Coleridge, *Biographia Literaria*, ed. George Watson (London: Dent, 1965), p. 91.

55. Hawking, *History of Time*, p. 166.

56. Ibid., pp. 40, 150–51, 39–42, 49–51, 136–48.

57. Gary Taubes, "Everything's Now Tied to Strings," *Discover*, November 1986, pp. 34–36, 42, 48–49. See also Hawking, *History of Time*, pp. 158–62.

58. Ifrah, *From One to Zero*, p. 432.

VI. THE ASCENDANCY OF ENGLISH

1. Robert McCrum, William Cran, and Robert MacNeil, *The Story of English* (New York: Viking, 1986), p. 60. I am indebted to this book for information used in the chapter. I do not, however, agree entirely with their apparent view that the varieties of English will continue to increase in number.

2. Ibid., p. 75.

3. Geoffrey Chaucer, *Works*, ed. F. N. Robinson, 2d ed. (Boston: Houghton Mifflin, 1957), p. 18.

4. Ibid., p. 479.

5. H. R. Loyn, *The Norman Conquest*, 3d ed. (London: Hutchinson, 1982) offers a good introduction to this subject.

6. McCrum, Cran, and MacNeil, *The Story of English*, pp. 93, 95.

7. Joseph Hunter, *New Illust. of the Life, etc. of Shakespeare* (London, 1845), 1.5. Cited from S. Austin Allibone, *A Critical Dictionary of English Literature and British and American Authors* (Philadelphia: Lippincott, 1874), p. 2006.

8. William Caxton, "Prologue to *Eneydos* (ca. 1490)," in *Caxton's Own Prose*, ed. N. F. Blake (London: André Deutsch, 1973), pp. 79–80.

9. McCrum, Cran, and MacNeil, *The Story of English*, pp. 19–20.

10. Daniel Defoe, "The True-Born Englishman," in *A Collection of English Poems 1660–1800*, ed. Ronald S. Crane (New York: Harper and Row, 1932), pp. 235–41.

11. Samuel L. Macey, *Clocks and the Cosmos: Time in Western Life and Thought* (Hamden, Conn.: Archon, 1980), p. 70.

12. Marjorie Nicolson, "The Early Stages of Cartesianism in England," *Studies in Philology* 26 (1929): 373.

13. Richard Foster Jones, "Science and English Prose Style in the Third Quarter of the Seventeenth Century," *PMLA* 45B (1930): 989ff.

14. Thomas Sprat, *The History of the Royal Society* (London: J. Martyn, 1667), pp. 413, 417.

15. Sprat, *Royal Society*, pp. 111, 62, 112.

16. Ibid., pp. 137, 113.

17. Jonathan Swift, *The Works*, ed. Herbert Davis (Oxford: Basil Blackwell, 1965), 11:185–86.

18. Macey, *Clocks and the Cosmos*, esp. p. 173.

19. John Dryden, *Works*, ed. Walter Scott (London: James Ballantyne, 1808), 15:231.

20. Alexander Pope, *Works*, ed. Whitwell Elwin and William John Courthope (London: John Murray, 1886), 10:554.

21. Alexander Pope, *Poems*, Twickenham Edition, ed. E. Audra and Aubrey Williams (London: Methuen, 1961), pp. 239–40.

22. Daniel Defoe, *An Essay upon Projects* (1697; reprint, Menston, U.K.: Scolar Press, 1969), pp. 227–52.

23. [Daniel Defoe], "Of the Trading Stile," in *The Compleat English Tradesman* (1727; reprint, New York: Augustus M. Kelley, 1969), 1.1:26–34.

24. Christiaan Huygens of Zulichem, *Horologium* (1658), trans. Ernest L. Edwardes, in *Antiquarian Horology* 7 (December 1970): 36.

25. Walter Charleton, *Physiologia Epicuro-Gassendo-Charltonia*, introduction by Robert Hugh Kargon (1654; reprint, New York: Johnson, 1966), pp. 76–77.

26. John Bonnycastle, *An Introduction to Astronomy* (London: J. Johnson, 1786), pp. 159–60.

27. Edward Tenner, "Cognitive Input Device in the Form of a Randomly

Accessible Instantaneous-Read-Out Batch-Processed Pigment-Saturated Laminous-Cellulose Hard-Copy Output Matrix," *Discover*, May 1986, pp. 58–61.

28. Roger D. Lund, "*Res et Verba*: Scriblerian Satire and the Fate of Language," *Bucknell Review* (1983): 63–80.

29. Samuel Daniel, "Musophilus; or, Defense of All Learning," (1602–1603)," in *Complete Works in Verse and Prose*, ed. Alexander B. Grosart (London: Spenser Society, 1885), 1:255.

30. McCrum, Cran, and MacNeil, *The Story of English*, pp. 182–83.

31. Thomas Hardy, *Tess of the d'Urbervilles*, ed. Scott Elledge (New York: Norton, 1965), p. 17.

32. McCrum, Cran, and MacNeil, *The Story of English*, p. 308.

33. Bertrand Marotte, "English Favored Language, Quebec Survey Finds," *Toronto Globe and Mail*, June 30, 1986.

34. For the contribution of blacks in particular, see McCrum, Cran, and MacNeil, *The Story of English*, pp. 195–233.

35. R. R. Palmer, *A History of the Modern World* (New York: Knopf, 1960), pp. 564–66; McCrum, Cran, and MacNeil, *The Story of English*, p. 266.

36. McCrum, Cran, and MacNeil, *The Story of English*, p. 264.

37. Ibid., p. 20.

38. Ibid.

VII. GREAT BRITAIN AND THE INDUSTRIAL REVOLUTION

1. I acknowledge with thanks the editor's permission to include material from my article "Work Study Before Taylor," *Work Study and Management Services* 18 (October 1974): 530–36. For an American view of European work study or management services—known in the United States as industrial engineering—see Serge A. Birn and Eric G. Brightford, "Industrial Engineering in Europe Today," *Work Study and Management Services* 14 (April 1970): 311–15 (reprinted from the *Journal of Industrial Engineering*).

2. John Dryden, *Works*, ed. Walter Scott (London: James Ballantyne, 1808), 11:214–16; Daniel Defoe, *A Journal of the Plague Year* (New York: New American Library, 1960), pp. 217–18.

3. Henry Atkinson, "Scientific Management," *Engineering and Industrial Management*, July 31, 1919, p. 137.

4. Ivor B. Hart, *The World of Leonardo da Vinci* (London: Macdonald, 1961), p. 170; and Mario Consiglio, "Leonardo da Vinci: The First IE?" *Industrial Engineering* 1 (January 1969): 71.

5. Lucian, *Selected Satires*, trans. Lionel Casson (New York: Norton, 1968), pp. 347, 349.

6. Plato, *Republic*, in *Great Dialogues*, trans. W. H. D. Rowse (New York:

New American Library, 1956), 369C–371B, passim. As late as the fourteenth century, Chaucer's wife of Bath still argues that, since practical experience produces perfection in particular workmen, her schooling by five husbands has made her the more perfect wife (Prologue to *Canterbury Tales*, lines 47–50).

7. Denis Diderot and Jean le Rond d'Alembert, eds., *Encyclopédie* (Neufchastel [Paris]: Samuel Faulche, 1751–65), 8:298–310.

8. Sir William Petty, *Another Essay in Political Arithmetic . . . 1682*, in *The Economic Writings*, ed. Charles Henry Hull (1899; reprint, New York: Augustus M. Kelley, 1963), 1:473.

9. Bernard Mandeville, *The Fable of the Bees; or, Private Vices, Public Benefits*, ed. F. B. Kaye (Oxford: Clarendon Press, 1924), 2:284; and Alexander Pope, *Works*, ed. Whitwell Elwin and William John Courthope (London: John Murray, 1886), 10:395–96.

10. Adam Smith, *An Inquiry into . . . the Wealth of Nations*, ed. Edwin Cannan (New York: Modern Library, 1937), p. 243.

11. Ferchault de Reamur, *Art de l'Épinglier . . . avec des additions de M. Duhamel du Monceau, et des remarques extraites de M. Perronet* [Paris, 1762], pp. 71–75, 59–70, plates 1–7, pp. 1–3, 43–44.

12. Henry Hamilton, *The English Brass and Copper Industries to 1800*, 2d ed. (London: Frank Cass, 1967), pp. 46–47, 255.

13. Ibid., pp. 346–49.

14. Ibid., p. 343; and William Cudworth, *Life and Correspondence of Abraham Sharp* (London: Samson Low, 1889), pp. 30–31.

15. H. W. Dickinson, *Matthew Boulton* (Cambridge: Cambridge University Press, 1936), pp. 133–34, 140.

16. H. Alan Lloyd, "Samuel Watson," *Antiquarian Horology* 1 (December 1954): 60–61.

17. R. M. Currie, *Work Study*, 2d ed. (London: Pitman, 1963), p. 2.

18. Sir Eric Roll, *An Early Experiment in Industrial Organization* (London: Frank Cass, 1968), pp. 169, 172, 293ff., 256–57, 247–49, 307–12, 217–19.

19. J. G. Crowther, *Scientists of the Industrial Revolution* (London: Cresset, 1962), pp. 152, 135; A. E. Musson and Eric Robinson, *Science and Technology in the Industrial Revolution* (Manchester: Manchester University Press, 1969), pp. 200ff.

20. Musson and Robinson, *Science and Technology*, p. 485.

21. Ibid., pp. 494–96.

22. M. W. Flinn, *Men of Iron* (Edinburgh: Edinburgh University Press, n.d.), pp. 190–93, 203.

23. Ibid., pp. 216–18, 220–21.

24. Roll, *Industrial Organization*, pp. 225–27.

25. Allardyce Nicoll, *Early Eighteenth Century Drama*, 3d ed., vol. 2 of *A History of the English Drama 1660–1900* (Cambridge: Cambridge University Press, 1965), p. 181.

26. M. W. Flinn, *Origins of the Industrial Revolution* ([London]: Longmans, 1966), p. 70; and Charles Babbage, *On the Economy of Machinery and Manufactures*, 4th ed., enlarged (1835; reprint, New York: Augustus M. Kelley, 1963), pp. 158–59. References are to the fourth edition unless otherwise stated.

27. Robert Owen, *Observations on the Effect of the Manufacturing System* (London: Richard and Arthur Taylor, 1815), pp. 6–7, 12–13.

28. Babbage, *Economy of Machinery*, 1st ed., pp. 229–30. This potentially controversial portion of the chapter appears to have been removed from the fourth edition.

29. Ibid., 4th ed., p. 117.

30. Frederick Winslow Taylor, *Scientific Management*, comprising *Shop Management*, *The Principles of Scientific Management*, and *Testimony Before the Special House Committee* (New York: Harper and Brothers, 1947); *Principles*, p. 129; and *Shop Management*, p. 99.

VIII. NORTH AMERICA AND THE WORLD

1. See however, S. J. Noel-Brown, "The History of Work Study," *Work Study and Management Services* 13 (December 1969): 816–19, 822; and C. A. Horn, "The Origins of Work Study in the UK," *Work Study and Management Services* 15 (April 1971): 260–67.

2. F. W. Taylor, "A Piece-Rate System," *Cassier's Magazine*, October 1895; H. L. Gantt, "A New System of Rewarding Machine Shop Labour," *Cassier's Magazine*, November 1902; also Taylor's *Testimony*, pp. 5–7, in Frederick Winslow Taylor, *Scientific Management*, comprising *Shop Management*, *The Principles of Scientific Management*, and *Testimony Before the Special House Committee* (New York: Harper and Brothers, 1947). Future references to Taylor's works will be made in the text. For an article questioning the veracity of some of Taylor's claims, see Charles D. Wrege and Amedeo G. Perroni, "Taylor's Pig-Tale," *Work Study and Management Services* 18 (November 1974): 564–75.

3. *Principles*, pp. 68–71; and *Testimony*, pp. 64–65. See *Shop Management*, pp. 111ff., for "functions of the planning department."

4. Taylor, *Shop Management*, pp. 148–78; and *Principles*, p. 67. See also Gantt, "New System."

5. Aldous Huxley, *Brave New World* (Harmondsworth, U.K.: Penguin, 1955), p. 18.

6. "Institute News," *Work Study and Management Services* 17 (October 1973): 726.

7. H. E. Kearsey, "The Bedaux Work Unit Method," *Work Study and Management Services* 14 (September 1970): 725.

8. Horn, "Origins of Work Study," p. 266.

9. F. B. Gilbreth and L. M. Gilbreth, *The Writings of the Gilbreths*, ed. William R. Spriegel and Clark E. Myers (Homewood, Ill.: Richard D. Irwin, 1953), p. 55.

10. S. Oliver, "An Outline of Ergonomics," *Work Study and Management Services* 14 (February 1970): 117–21.

11. For MTM in connection with tape data, see F. Evans, "Tape Data Analysis," *Work Study and Management Services* 13 (December 1969): 835–37; for Basic Work Data, see "BWD Goes Metric," *Work Study and Management Services* 14 (June 1970): 484–86.

12. "The British Work-Measurement Data Foundation," *Work Study and Management Services* 14 (May 1970): 376–78.

13. James E. Kelley, Jr., "CPM: Present and Future," in *New Horizons in Industrial Engineering*, ed. Seymour M. Selig and Morton Ettelstein (Baltimore, Md.: Spartan, 1963), pp. 1–6.

14. G. T. Brown, "Cybernetics and Work Study," *Work Study and Management Services* 17 (April 1973): 228–32.

15. Anthony Trollope, *An Autobiography* (London: Oxford University Press, 1950), pp. 118–19, 271–72.

16. R. M. Currie, *Work Study*, 2d ed. (London: Pitman, 1973), p. 39.

17. See my articles: "On Dividing the Loot: The Delegation of Power," *Yale Review* 61 (March 1972): 396–406; and "Who Cares What Happens Thirty Years from Now?" *Winnipeg Free Press*, February 10, 1972.

18. David Halberstam, *The Reckoning* (New York: William Morrow, 1986), pp. 79–81.

19. H. W. Dickinson, *Matthew Boulton* (Cambridge: Cambridge Unversity Press, 1936), pp. 58–59; Halberstam, *The Reckoning*, p. 74.

20. Cited from James Sexton, "*Brave New World* and the Rationalization of Industry," *English Studies in Canada* 12 (December 1986): 429.

21. Samuel L. Macey, "The Role of Clocks and Time in Dystopias: Zamyatin's *We* and Huxley's *Brave New World*," in *Explorations: Essays in Comparative Literature*, ed. Makoto Ueda (Lanham, Md.: University Press of America, 1986), pp. 29–33.

22. Halberstam, *The Reckoning*, pp. 87–88.

23. John Dryden, "Preface to the Fables," in *Works*, ed. Walter Scott (London: James Ballantyne, 1808), 11:214–16; [Daniel Defoe], *A Plan of the English Commerce* ([1730]; reprint, Augustus M. Kelley, 1967), p. 299.

24. Daniel Defoe, *Moll Flanders* (New York: Random House, 1950), pp. 114, 210.
25. Halberstam, *The Reckoning*, p. 82.
26. Peter Spry-Leverton and Peter Kornicki, *Japan* (London: Michael O'Mara, 1987), p. 176; "The Next Japans," *The Economist*, July 30, 1988, p. 13.
27. Walt Whitman Rostow, "Industrialized Nations Facing Third World Challenge," *Victoria Times-Colonist*, September 7, 1986.
28. "World Watch Production Increases: Total Market Value Declines," *Horological Journal* 129 (May 1987): 28–29; also 131 (June 1989): 9.

IX. RETAIL DISTRIBUTION AND FINANCE

1. Meg Whittemore, "The Great Franchise Boom," *Nation's Business*, September 1984, pp. 20–24.
2. Evan E. Anderson, "The Growth and Performance of Franchise Systems: Company Versus Franchisee Ownership," *Journal of Economics and Business* 36 (1984): 421–31.
3. Cited from Meg Whittemore, "Franchising's Future," *Nation's Business*, February 1986, 47–50. "Report on Entrepreneurship and Franchising," *Toronto Globe and Mail*, September 5, 1988, C1, C7, C15.
4. "Franchising Goes International," *Venture*, July 1985, pp. 98–101. See also Robert E. Bond, *The Source Book of Franchise Opportunities* (Irwin: Dow Jones, 1985), p. 3.
5. Daniel Defoe, *A Journal of the Plague Year* (New York: Signet, 1960), p. 18. For a useful discussion of the present growth of service industries, see James Brian Quinn, Jordan J. Baruch, and Penny Cushman Paquette, "Technology in Services," *Scientific American*, December 1987, pp. 50–58.
6. "Franchising Goes International," p. 98.
7. Doris Walsh, "Bite the Wax Tadpole," *American Demographics*, March 1986, p. 48.
8. George Chandler, *Four Centuries of Banking* (London: B. T. Batsford, 1964), 1:41–42.
9. Robin D. Gwynn, *Huguenot Heritage* (London: Routledge and Kegan Paul, 1985), pp. 60–61, 152–59.
10. Samuel L. Macey, "The Time Schemes in *Moll Flanders*," *Notes and Queries* 214 (September 1969): 336–37.
11. Samuel L. Macey, *Money and the Novel: Mercenary Motivation in Defoe and His Immediate Successors* (Victoria, B.C.: Sono Nis, 1983), pp. 87–98, 151–57.

12. Sidney Homer, *A History of Interest Rates* (New Brunswick, N.J.: Rutgers University Press, 1977), p. 124.

13. [Daniel Defoe], *A Plan of the English Commerce* ([1730]; reprint, New York: Augustus M. Kelley, 1967), pp. 80–84.

14. Philip Beresford and John Cassidy, "The Big Year of the Small Shareholder," *Sunday Times* (London), December 21, 1986.

15. John Kohut, "Land of the Rising $um," *Toronto Globe and Mail*, April 25, 1987. See also "US Debt Seen Topping $2 Trillion" (based on an Associated Press report), *Toronto Globe and Mail*, July 4, 1988.

16. Ibid.

17. David Halberstam, *The Reckoning* (New York: William Morrow, 1986), p. 53.

X. LABOR AND THE INDIVIDUAL

1. Cited from E. P. Thompson, "Time, Work-Discipline, and Industrial Capitalism," *Past and Present* 38 (December 1967): 61.

2. Cited from Thompson, "Time," p. 61.

3. Stephen Duck, "The Thresher's Labour," in *The New Oxford Book of Eighteenth Century Verse*, ed. Roger Lonsdale (New York: Oxford University Press, 1984), p. 225.

4. Samuel Smiles, *Industrial Biography: Iron Workers and Tool Makers* (1863; reprint, Newton Abbot, U.K.: David and Charles, 1967), pp. 180–81.

5. Cited from Neil McKendrick, "Josiah Wedgwood and Factory Discipline," *Historical Journal* 4 (1961): 34.

6. Andrew Ure, *The Philosophy of Manufactures; or, An Exposition of the Scientific, Moral, and Commercial Economy of the Factory System of Great Britain* (1835; reprint, London: Frank Cass, 1967), p. 20.

7. See Thompson, "Time," pp. 72–75.

8. Ure, *The Philosophy of Manufactures*, pp. 20–23.

9. Ibid., p. 41.

10. Sir James P. Kay-Shuttleworth, cited from Frank Smith, *The Life and Works of Sir James Kay-Shuttleworth* (Bath: Cedric Chivers, 1974), pp. 21–23.

11. Charles Babbage, *On the Economy of Machinery and Manufactures* (London: Charles Knight, 1832), pp. 278–79.

12. A. E. Musson and Eric Robinson, *Science and Technology in the Industrial Revolution* (Manchester: Manchester University Press, 1969), pp. 435, 457, 436–37, 24, 438, 143; Josiah Wedgwood, *Selected Letters*, ed. Ann Finer and George Savage (London: Cory, Adams and Mackay, 1965), plate

10; and J. G. Crowther, *Scientists of the Industrial Revolution* (London: Cresset Press, 1962), p. 130.

13. James Sexton, "*Brave New World* and the Rationalization of Industry," *English Studies in Canada* 12 (December 1986): 426–27.

14. Cited from Gilbert J. French, *Life and Times of Samuel Crompton* (Bath: Adams and Dart, 1970), pp. 261–63.

15. George Oldfield, *The Hungry Forties: Life Under the Bread Tax* (London: T. Fisher Unwin, 1904), pp. 195–98.

16. Cited from Edwin Hodder, *The Life and Work of the Seventh Earl of Shaftesbury, K.G.* (London: Cassell, 1892), p. 314.

17. Cited from Humphrey Jennings, *Pandaemonium: The Coming of the Machine as Seen by Contemporary Observers, 1660–1886*, ed. Mary-Lou Jennings and Charles Madge (New York: Free Press, 1985), pp. 219–20.

18. Cited from Jennings, *Pandaemonium*, pp. 316–17.

19. Cited from Jennings, *Pandaemonium*, pp. 21–22.

20. James Burke, *Connections* (Boston: Little, Brown, 1978), p. 155.

21. [Daniel Defoe], *A Plan of the English Commerce* ([1730]; reprint, New York: Augustus M. Kelley, 1967), pp. 243–44.

22. Thompson, "Time," pp. 70–84.

23. Robert Southey, *Journal of a Tour in Scotland in 1819* (London: John Murray, 1929), pp. 263–65.

24. Joseph Farington, *The Farington Diary*, ed. James Greig (London: Hutchinson, [1922]), 1:314.

25. Paul Mantoux, *The Industrial Revolution in the Eighteenth Century* (London: Jonathan Cape, 1961), pp. 216–17, 228–32.

26. Cited from Jennings, *Pandaemonium*, pp. 76–77.

27. Sir William Napier, *The Life and Opinions of General Sir Charles James Napier* (London: John Murray, 1857), pp. 41–42.

28. Friedrich Engels, *The Conditions of the Working Class in England in 1844*, trans. Florence Kelley Wischnewetzky (London: Swan Sonnenschein, 1892), pp. 41–42.

29. Cited from Jennings, *Pandaemonium*, pp. 179–80.

30. David Halberstam, *The Reckoning* (New York: William Morrow, 1986), pp. 340–43.

31. Ibid., pp. 608, 186–87.

32. Charles Babbage, cited in James Phillips Kay[-Shuttleworth], *The Moral and Physical Condition of the Working Classes* (1832; reprint, Manchester: E. J. Morton, 1969), pp. 108–9.

33. Associated Press, "Inventory System Worsens Strike," *Victoria Times-Colonist*, November 21, 1986.

34. Halberstam, *The Reckoning*, pp. 620–22.

35. Ibid., pp. 632–36. But see *The Economist*, May 20, 1989, p. 58ff.

36. Samuel L. Macey, "On Dividing the Loot: The Delegation of Power," *Yale Review* 59 (March 1972): 396–406.

37. "U.S. Postal Service Pinpoints Our Problems," *Victoria Times-Colonist*, July 13, 1987.

XI. HUMAN EQUALITY

1. Alexander Pope, *An Essay on Man*, Twickenham Edition, ed. Maynard Mack (London: Methuen, 1964), pp. 44–49.

2. Geoffrey Chaucer, *Poetry*, ed. E. T. Donaldson (New York: Ronald Press, 1958), pp. 894–95.

3. Geoffrey Chaucer, *Works*, ed. F. N. Robinson, 2d ed. (Boston: Houghton Mifflin, 1957), p. 20 ("General Prologue," lines 361–78).

4. R. R. Palmer, *A History of the Modern World*, 2d ed. (New York: Knopf, 1960), pp. 155, 162–63.

5. James Burke, *The Day the Universe Changed* (London: British Broadcasting Corporation, 1985), p. 165.

6. George Macaulay Trevelyan, *British History in the Nineteenth Century and After, 1782–1919* (New York: Longmans, Green, 1945), p. 189.

7. J. L. Hammond and Barbara Hammond, *The Town Labourer, 1660–1832: The New Civilization* (London: Longmans, Green, 1925), pp. 90–91.

8. Palmer, *History*, p. 425.

9. Thomas Walkom, "Japanese Women Work More, Earn Less," *Toronto Globe and Mail*, October 9, 1986.

10. Daniel Defoe, *The Compleat English Tradesman* (1727; reprint, New York: Augustus M. Kelley, 1969), 1.1:114–15.

11. J. M. Roberts, *The Triumph of the West* (London: British Broadcasting Corporation, 1985), p. 19.

12. Samuel L. Macey, *Clocks and the Cosmos: Time in Western Life and Thought* (Hamden, Conn.: Archon, 1980), pp. 74–77.

13. See, for example, the Bruntland report of the United Nations' World Commission on Environment and Development, as discussed in Michael Keating's "Pollution Risks 'Survival of Life,'" *Toronto Globe and Mail*, April 27, 1987.

CONCLUSION

1. Samuel L. Macey, *Clocks and the Cosmos: Time in Western Life and Thought* (Hamden, Conn.: Archon, 1980), p. 194.

2. René Descartes, *Discourse on Method*, trans. F. E. Sutcliffe (Harmondsworth, U.K.: Penguin, 1968), pp. 73–76, and *Philosophical Letters*, trans. Anthony Kenny (Oxford: Clarendon Press, 1970), pp. 53–54, 207–8, 244.

Index

Académie Française, 116–17, 130
Academy of Sciences, French, 75
Adams, John Quincy, 77
Addison, Joseph, 118
Adelard of Bath, 95
Adonis, 43
Ahriman, 44
Aidan, Saint, 112
Alexander, Samuel, 105
Alfred, King, 7, 112–13
Alhazen, 96, 108
Alkon, Paul, 32, 46–47
Alphonso the Wise, 95
American System of Manufacture, 204, 237. *See also* Industrial revolution
Archimedes, 99
Aristotle, 30, 96, 218, 233
Arkwright, Richard, 208–11, 219, 237
Augustine of Hippo, Saint, 112
Augustus Caesar, 29, 45
Aurelian, 26
Austen, Jane, 48, 189

Babbage, Charles, 144, 149–54, 165, 203, 237

Bacon, Francis, 53, 74, 111, 236
Bank of England, 188–90, 196
Banking and finance, 188–94; corporate, 174, 195–96; modern corporate managers, 193–94; government, 194–97; personal, 195
Barth, Carl G., 156, 160
Bartusiak, Marcia, 104
Bedaux, Charles Eugene, 149, 154
Bede, the Venerable, 45, 85
Benedict, Saint, 9
Bentley, Thomas, 210
Beresford, Philip, 193
Berkeley, George, 78, 108
Bible, 49, 127, 229; Old Testament, 31, 42, 44; New Testament, 44
Bickerman, Elias Joseph, 4–5, 8, 45
Big bang theory, 110
Black hole, 109
Blake, William, 23, 149
Bligh, William, 12
Boethius, Anicius Manlius Severinus, 112
Boileau-Despréaux, Nicolas, 120
Bonnycastle, John, 123–24
Boorstin, Daniel J., 54

Boswell, James, 78
Boulton, Matthew, 145–46, 169, 187, 202, 237
Boyle, Robert, 54, 118–19, 143
Bradley, Langley, 124
Brahe, Tycho, 11, 97
Brahma, 43
Brecht, Bertolt, 107
Bridgwater, Francis Egerton, third duke of, 211
Bronowski, Jacob, 96–98, 106
Browne, Thomas, 118
Brunelleschi, Filippo, 96
Bruton, Eric, 12
Buddha, 87
Buffon, Georges Louis Leclerc, comte de, 50, 53–54, 233
Bullion and currency, 73–74; dirhem 73; esterling or sterling, 73; moneyer's pound, 73; silver penny, 73; tower pound, 73
Burchfield, Robert, 129
Burgi, Jost, 11
Burke, James, 59, 97, 103, 207, 221
Burns, Robert, 134
Burritt, Elihu, 206
Burton, Robert, 118

Cairo Museum, 6
Calendars: Julian, 28–29, 47, 232; Aztec, 30–31; Baha'i, 30; Gregorian, 30, 32–34, 38, 76, 233, 239; Mayan, 30–31; British, 32–33; Scottish, 33; French Republican, 34–37; Soviet, 36–38; Egyptian solar, 39–40
Calvin, John, 188
Cardano, Jeronimo, 95
Carson, Rachel, 229
Carter, Brandon, 107
Cassidy, John, 193

Cassini, Giovanni Domenico, 13
Caxton, William, 115–16
Censorinus, 27–28
Ceram, C. W. [Kurt W. Marek], 30–31
Chambers, Ephraim, 71, 142, 144
Charlemagne, 45, 66, 91
Charles II, 19, 111, 126
Charleton, Walter, 122–23
Chartists, 211
Chaucer, Geoffrey, 4, 9, 17, 53, 113–15, 122, 220, 254 (n. 6)
Chuang-Tzu, 43
Christmas, 38
Chronology, 41–61, 233–34; Sothic period, 27; Judaic, 31; Roman Catholic Holy Year, 31; Bible, 41–42, 48, 58, 60; Greek, 41; Roman, 41, 44; Mayan, 43; Greek Olympiads, 44–45; Easter, 45; Christian, or Incarnation, Era, 45–46; Chinese, 48–49, 51; Christian, 48, 51–52; Islamic, 51; present model of the universe, 51
Clark, Samuel, 42
Clavius, Christopher, 30
Clepsydrae (water clocks), 7, 140–42. *See also* Clocks, Sandglasses, Mechanical clocks
Clocks: atomic, 20; hydrogen masers, 20–21. *See also* Mechanical clocks
Club of Rome, 229
Colbert, Jean Baptiste, 221
Coleridge, Samuel Taylor, 56, 108
Columbus, Christopher, 13
Communism, 211–12, 215
Condorcet, Antoine Nicolas, marquis de, 50, 53–54, 233
Cook, James, 12
Cooper, Anthony Ashley, seventh earl of Shaftesbury, 205–6

Copernicus, Nicholas, 11, 29–30, 93, 233
Coster, Salomon, 12
Coulomb, Charles Augustin de, 151
Coulton, George Gordon, 70
Cran, William, 131
Cronus, 5, 43
Crowley, Ambrose III, 147–48, 208, 237
Currie, Russel Mackenzie, 146, 150, 155, 165
Cuvier, Georges, 56

Dafydd ap Gwilym, 22
Daimyō Clock Museum, 8
Daniel, Samuel, 127
Darwin, Charles, 23, 50, 55–57, 59, 219, 233
David I, king of Scotland, 69–70
Dawkins, Richard, 101–2, 106
Decimal system: millimeter, 79. See also Metric system
Defoe, Daniel, 14, 46–48, 58, 84, 116–18, 121–22, 171–75, 187–90, 206–8, 226, 241
Deming, W. Edward, 214–15
Democritus, 77
Derham, William, 124
Descartes, René, 23, 53, 74, 95–96, 108, 120, 142, 229, 242
Dickens, Charles, 165, 210
Diderot, Denis, 124, 142–43, 163, 237
Diogenes Laërtius, 77–78
Dionysius Exiguus, 45
Disraeli, Benjamin, 222–23
Divisions of the day, 3–4, 231–32; Achanese, 3; at Cairo, 3; Javanese, 3; Madagascar, 3; Malay, 3; unequal, temporal, or variable hours, 3–4, 8–11; Babylonian, 4–5; Chinese, 4–5, 21; Egyptian, 4–6; European, 4; Greek, 4–5; Hindu, 4–5; Japanese, 4–5, 8, 21; at London, 4; "tides," 5; watches, 5; Homer, 5; Judaic, 5; Old Testament, 5; Roman, 5–6; Saxon, 5; British, 8; equal hours, 8, 21; the hour, 8; "hour," 8; canonical hours, 9; *horarium*, 9; Greenwich mean time, 18–19; International Meridian Conference, 19; standard time, 19; French, 22; Mesopotamian, 21–22
Divisions of the hour, 10, 17; atom, 79–81; minute, 79–80; moment, 79–81; point, 79; second, 79
Divisions of the second, 22. See also Divisions of the hour
Divisions of the year, 25, 232–33; Assyrian, 25; Babylonian, 25; Central American, 25; Christian, 25–26; Egyptian, 25–28; Inca, 25; Judaic, 25–26, 37–39; Roman, 25–26; Teuton, 25; West African, 25; lunar month, 26–27; Samoan, 26; seven-day week, 26, 39–40, 77; Egyptian solar year, 27–28; the seasons, 27; solar year, 27; synodic month, 27; Christian, 37–40; Islamic, 37
Donaldson, Ethelbert Talbot, 220
Donne, John, 15
Doppler effect, 109
Dow Jones Industrial Average, 192
Dowd, Charles F., 19
Dryden, John, 114, 118, 120–21, 139–40, 171
Duck, Stephen, 202
Durant, William Crapo, 169–70
Dürer, Albrecht, 97

Echolocation, 101–5
Edinburgh, Philip Mountbatten, duke of, 162
Edward I, 71
Edward III, 72
Einstein, Albert, 51, 98, 100, 105–7, 109–10, 196, 234
Elizabeth I, 72–74
Engels, Friedrich, 50, 59, 211–12
English language, 112, 235–36; Angles, 112–13; the beginnings, 112–14; Old English, 112–13; Saxons, 112–13; Danes, 113; French influence, 113; Normans, 113; Latin influence, 113–14; Middle English, 113–14; Vikings, 113; Greek influence, 114; American influence, 116, 126–27, 130–32; vocabulary, 116; influence of science and technology, 119–24, 133; eliminating jargon, 125–27; internationalization and standardization, 127–30; influence of radio and television, 129–30, 132–34; reaction to internationalization, 130–35; British Commonwealth, 132–33; influence on non-English-speaking countries, 134–35
Equation of time, 14–15, 244 (n. 24)
Eratosthenes, 45
Eschatology, 42–44; Christian, 42, 44; Judaic, 42, 44; Islamic, 42, 44
Ethelred the Unready, 66
Euclid, 95

Fairbairn, William, 147
Farington, Joseph, 208
Father Time, 242
Fichte, Johann Gottlieb, 78, 108

Fielding, Henry, 48, 221
Flamsteed, John, 99
Fleming, Sir Sandford, 19
Flinn, Michael Walter, 147–48
Ford, Henry, 158, 168–73, 179, 205, 214, 238–39. *See under* Manufacturers: Ford Motor Company
Franchising: *See* Wholesale and retail distribution
Franklin, Benjamin, 23, 236
Fraser, J. T., 105–6
Free trade: Britain, 178–79; Canada, 178; European Community, 178–79; United States, 178–79; Greece, 179; Italy, 179; Portugal, 179; Spain, 179
Freud, Sigmund, 51, 233
Friedmann, Alexander, 109
Frisius, Gemma, 12
Froissart, Jean, 219–20
Fromanteel, Ahasuerus, 12

Gaia, 5, 43
Galen, 96
Galilei, Vincenzo, 11
Galileo Galilei, 11, 53, 98–100, 233
Gama, Vasco da, 13
Gandhi, Indira, 133
Gandhi, Rajiv, 133
Gantt, Henry L., 155, 159–60
Gerard of Cremona, 95
Ghiberti, Lorenzo, 96
Gilbreth, Frank Bunker, 162–64
Gilbreth, Lillian M., 162–64
Glanvill, Joseph, 118
Global village, 3, 19, 194–97, 232
Goethe, Johann Wolfgang von, 24
Goldsmith, Oliver, 149, 175
Gorbachev, Mikhail, 178
Graunt, John, 57–58

INDEX

Great Chain of Being, 218–19
Greenwich time, 13
Gregory XIII, Pope, 30, 36
Griffin, Donald, 101–2
Gromyko, Andrei, 178

Haeckel, Ernst, 59
Halberstam, David, 168, 170–71, 173, 195–96, 212–13, 215
Hall, Basil, 18
Halley, Edmund, 58, 99, 115
Halsey, F. A., 155
Hamilton, Henry, 145
Hammond, Barbara, 222
Hammond, John Lawrence, 222
Hardy, Thomas, 128, 134
Hargreaves, James, 210
Harrison, John, 12, 103, 231
Hartley, David, 142
Harvard-Smithsonian Center for Astrophysics, 20–21
Hauptmann, Gerhart, 166
Hawking, Stephen, 106, 108–10
Hazlitt, William, 23
Hearne, Thomas, 46–47
Hecker, J. F. K., 220
Hegel, Georg Wilhelm Friedrich, 50
Henley, Walter de, 140
Henriques, Basil, 216–17
Henry I, 69, 113
Henry VII, 72–73
Herod, 45
Herodotus, 4
Hesiod, 43
Hill, Rowland, 18
Hitler, Adolf, 191
Hobbes, Thomas, 166
Hogarth, William, 32, 182
Holy Sepulchre, 38
Homer, Sidney, 189–90
Hooke, Robert, 124

Horace, 78, 120
Horloges, 7. *See also* Clocks
Horological revolution, British, 12, 23–24, 47, 117, 125, 233
Howse, Derek, 15, 18
Hubble, Edwin, 109
Hugo, Victor, 24
Huguenots, 188–89
Human franchise, 223–27; and democratization of shareholdings, 192–94. *See also* Banking and finance, Rationalizing human beings
Hume, David, 56–57
Hunter, Joseph, 115
Hutton, James, 54
Huygens, Christiaan, 11–12, 122
Huxley, Aldous, 158, 170, 205
Huxley, Thomas Henry, 59

Ifrah, Georges, 31, 83–86, 88, 92, 94, 110
Industrial revolution, 12, 17, 236; British, 204–205. *See also* American system of manufacture

James I, 145
Jerome, Chauncey, 237
John, king of England, 66–67
Johnson, Samuel, 78, 117, 120, 205, 219
Johnstone, William D., 79
Jones, Richard Foster, 53, 118
Jones, R. V., 103
Jones, William Powell, 56
Julius Caesar, 28–29, 40, 45

Kaiser, Henry, 172, 213
Kant, Immanuel, 53, 108
Kay, John, 210
Kay-Shuttleworth, James, 203, 214

Kearsey, H. E., 161
Kennedy, R. J., 100
Kepler, Johannes, 11, 97
Kisch, Bruno, 66
Kohut, John, 194
Koran, 44, 51
Kornicki, Peter, 174
Kwarizmi, al-, 95

Labor-management relations, 181; American, 169; British, North American, Japanese, 171–72; British, West German, Japanese, American, 172–74; British, American, 174–77; South Korea, Taiwan, Singapore, Hong Kong, 174. *See also* Union-management relations
Lamb, Charles, 23
Langland, William, 220
Large, Robert, 115
Larin, Comrade, 36
Latitude, 12. *See also* Navigation, Longitude
Leibniz, Gottfried Wilhelm, 23, 58, 96–98
Lenin, 36
Lentulus, Cossus Cornelius, 45
Lessing, Gotthold Ephraim, 97
Ligachev, Yegor, 178
Lightfoot, John, 49
Linnaeus, Carolus, 56
Locke, John, 47–49, 53, 142, 206
Longitude, 12; prize of twenty thousand pounds, 14. *See also* Navigation, Latitude
Lorentz, Hendrick Antoon, 105
Louis XIV, 220–21
Lowe, Robert, Viscount Sherbrooke, 223

Lucian, 142
Lyell, Charles, 50, 55, 59, 233

Machiavelli, Niccolo, 166
McCrum, Robert, 131
MacNeil, Robert, 131
Malthus, Thomas Robert, 57–58, 219, 226
Mandeville, Bernard, 143
Mansell-White, J., 160
Manufacturers: Soho Works, 154; Bethlehem Steel, 155; Midvale Steel Company, 155–56, 159; Huntley and Palmers, 162; Imperial Chemical Industries, 162, 170; J. Lyons and Company, 162; Ford Motor Company, 168–71, 214; General Motors, 169–70, 172, 213, 215; Toyota, 171; Citizen Watch Company, 179–80, 238; Standard Oil, 213–14; Nissan, 215–16
Markham, G., 201
Marlowe, Christopher, 115
Marx, Karl, 36, 50, 59, 211–12, 233
Maskelyne, Nevil, 13
Mechanical clocks, 7; Japanese, 7; spring drive, 7; verge-and-foliot escapement, 7, 10; Salisbury Cathedral, 9; Wells Cathedral, 9; pendulum escapement, 11; marine chronometer, 12; spring-balance escapement, 12; pendulum, 14; spring-balance watches, 14. *See also* Clocks
Menninger, Karl, 82–86, 88, 91–92
Mercator, Gerardus, 12
Method Study. *See* Time and motion study
Methuselah, 43

INDEX

Metric system, 22; United States, 33, 76–77; meter, 74–77; centimeter, 75; decimeter, 75; gram, 75; Great Britain, 76–77; kilogram, 76; Canada, 77; liter, 77; millimeter, 79. *See also* Weights and measures
Michelson, Albert Abraham, 100
Miller, Arthur, 182
Minkowski, Hermann, 105
Mithraists, 26
Mohammed, 51
Mond, Alfred Moritz, 170
More, Thomas, 10
Morgan, Lewis Henry, 50, 233
Morley, Edward W., 100
Mouton, Gabriel, 74
Mumford, Lewis, 8–9
Murdock, William, 204–5
Musson, Albert Edward, 147

Naisbitt, John, 186
Napier, Charles, 211
Napier, John, 90
Napoleon, Bonaparte, 35, 76
Nasmyth, James, 147
Navigation, 20. *See also* Latitude, Longitude
Nehru, Jawaharlal, 132–33
Newcomen, Thomas, 17
Newton, Isaac, 23, 42, 53–54, 57, 96–99, 120
Nicoll, Allardyce, 148, 206
Nicolson, Marjorie, 118
Nilsson, Martin P., 3, 26, 81
Number systems, 235; Hindu, 43; Hindu-Arabic numerals, 43, 86–89, 95, 110; Mesopotamian, 43, 94; early systems, 82–84; Sumerian, 83, 92, 94; finger counting, 84–86, 92; large numbers, 86; tally sticks, 86; Judaic, 87; Minoan, 87; Roman, 87; decimal, 90; Mayan, 90; vigesimal, 90; duodecimal, 91–93; Assyrian, 92; Babylonian, 92–93, 95; sexagesimal, 92–93; Elam, 94; Greek, 94–96

Offa, king of Mercia, 73
Ohrmazd, 44
O'Keefe, John, 73
Oldenburg, Henry, 52
Oldfield, George, 205
O'Neil, William Matthew, 45
O'Neill, Eugene, 182
OPEC, 172–74, 213
Opie, Iona, 38
Opie, Peter, 38
Orientation, 39. *See also* Divisions of the year
Otokar, king of Bohemia, 66
Owen, Robert, 148–49, 208
Ovid, 86

Palmer, Robert Roswell, 220, 224
Pankhurst, Emmeline, 224
Parthenon, 39
Paterson, William, 188
Paul, Alice, 224
Paul, Lewis, 205
Pelham, Henry, 190
Penrose, Roger, 109–10
Perronet, Jean (Rodolphe), 144, 152, 163
Perry, Matthew Calbraith, 4
Persephone, 43
Petty, William, 143, 206–7
Picard, Jean, 74
Piso, Lucius Calpurnius, 45

Pius XII, Pope, 59
Plato, 78, 83, 142, 163, 218
Plautus, 22–23
Plotinus, 218
Political parties, 176–78; French National Assembly, 176–77; Democrats, 177; Republicans, 177; Communists, 178
Polybius, 45
Pope, Alexander, 17, 79, 121, 143, 194, 219
Pretextatus, 66
Protagoras, 77–79
Ptolemy, 30, 93–95, 107, 233
Purchas, Samuel, 13

Quare, Daniel, 244 (n. 24)

Railways: British Columbia, 19; Canada, 19; Canadian Pacific, 19; United States, 19
Raleigh, Walter, 115
Rationalizing human beings, 218–30; Elementary Education Act of 1870, 128, 223; Reform Bills of 1832 and 1867, 222–23; Western technology and the franchise, 223–30, 239; Third World, 240. *See also* Rationalizing the workers
Rationalizing the workers, 201–2; urban labor, 202–6; slave labor and wage labor, 206–11, 223–24; the union movement, 210–17. *See also* Rationalizing human beings
Ray, John, 23, 56
Reactions against rationalization, 22–24, 35–40, 58–60, 77, 110–11, 130–35, 148–50, 165–75 passim, 202–17 passim, 239–40
Red-shift, 109

Remington, Eliphalet, 165
Retailers and service companies: F. W. Woolworth, 182–83, 238; Boots (the chemists), 183; J. Lyons and Company, 183; Marks and Spencer, 183, 238; W. H. Smith, 183; Amway, 185; Baskin-Robbins, 186; Century 21, 186; McDonald's Corporation, 186; Merry Maids, 186; Ramada Inns, 186; Sheraton Inns, 186; Coca-Cola, 187; Ford Motor Company, 192; American Telephone and Telegraph Company, 214; Bell Telephone Company, 214; Timex, 214; Canada Post, 216; U.S. Postal Service, 216, 238. *See also* Manufacturers; Ford, Henry; Woolworth, Frank Winfield
Rheingold, Erasmus, 30
Richard II, 113–14
Richardson, Samuel, 23, 48
Richelieu, 117
Roberts, John M., 227–28
Robinson, Eric, 147
Rockefeller, John Davison, 213–14
Roemer, Ole, 99–100
Roland, 43
Roll, Eric, 148
Romme, Charles-Gilbert, 35
Romulus, 28
Roosevelt, Franklin D., 213
Rothman, Tony, 107–8
Royal Navy, 5–6, 129
Royal Philosophical Society, 57
Royal Society, 52, 75, 111, 119, 139, 221, 236
Rush, Philip, 68, 73
Russell, Bertrand, 83
Russell, John, fourth duke of Bedford, 205

INDEX 271

Sandglasses, 7, 140
Santorio, Santorio (Sanctorius), 99
Savery, Thomas, 17
Schliemann, Heinrich, 50–51, 233
Schmandt-Besserat, Denise, 83
Science Museum, London, 6, 8, 140
Scipione del Ferro, 95
Scopes, John T., 59–60
Sexton, James, 170
Shadwell, Thomas, 166
Shakespeare, William, 8, 10, 17, 41, 115, 127
Sholes, Christopher Latham, 165
Shovel, Sir Cloudsley, 14
Sidney, Sir Philip, 114–15
Silverberg, Robert, 43, 49, 60
Sinclair, Upton, 131
Sirius, 27
Sloan, Alfred P., Jr., 170
Smith, Adam, 50, 144, 233
Smith, John, 124
Smith, William, 50, 54–55, 233
Socrates, 78
Sommers, John, baron, 135
Sosigenes of Alexandria, 29
South Sea Bubble, 192
Southey, Robert, 208
Space-time (also distance-time), 98–107
Spenser, Herbert, 59
Spinoza, Benedict (Baruch) de, 53
Sprat, Thomas, 118–19
Spry-Leverton, Peter, 174
Stalin, Joseph, 36–38
Standard Time, 232. *See also* Divisions of the day
Steno, Nicholas, 52–53
Sterne, Laurence, 146, 206
Stevin, Simon, 90
Stock Exchange: London, 192; New York, 193

Sundials, 6, 15
Swift, Jonathan, 24, 39, 118–19, 122, 126, 135, 206, 241
Sylvester II, Pope, 88
Szamosi, Geza, 98–99

Tacitus, 171
Talleyrand-Perigord, Charles Maurice de, 75
Taubes, Gary, 110
Taylor, Frederick Winslow, 146, 151, 154–61, 163–64, 166–67, 170–71, 205, 213, 238
Tenner, Edward, 125
Terry, Eli, 204, 237
Thatcher, Margaret, 193, 216
Thompson, Edward Palmer, 208
Thompson, Robert, 67
Thompson, Sanford E., 159
Thomsen, Christian Jürgensen, 50, 233
Timaeus, 44–45
Time and method study, 95, 149. *See also* Time and motion study
Time and motion study, 139–42; division of labor, 142–46; piece work and organization, 146–48; the influence on workers, 148–51; the influence of Babbage, 151–53; relationship with management, 153–54; Frederick Winslow Taylor, 155–61; Charles Eugene Bedaux, 161–62; American Institute of Industrial Engineers, 162, 165; Frank Bunker Gilbreth, 162–64; Lillian M. Gilbreth, 162–64; Institute of Management Services, 162, 165; CPM (Critical Path Method), 164; cybernetics, 164–65; ergonomics, 164; games techniques, 164; operational

Time and motion study (cont'd)
research, 164; PERT (Program Evaluation and Review Technique), 164; management, 167–68; Henry Ford, 168–71; time measurement and human beings, 179–80. See also Time and method study and entries under the names of individual practitioners
Tippett, L. H. C., 162
Titans, 43
Todhunter, Isaac, 92
Tompion, Thomas, 15, 124, 143, 179, 237
Toyoda, Eiji, 171
Trevelyan, George Macaulay, 222
Trevisa, John, 9–10, 17, 80
Trollope, Anthony, 165
Turkish influence on Western navigation, 13. See also Navigation

Union-management relations, 237. See also Labor-management relations
United Auto Workers, 212–13, 215
Uranus, 5, 43
Ure, Andrew, 202–3
Ussher, James, 41–42, 49–52, 60, 77, 106, 233–34, 240

Vancouver, George, 12
Vessot, Robert, 106
Vieta, François, 95
Vinci, Leonardo da, 140, 157
Vitruvius (Marcus Vitruvius Pollio), 6
Voltaire, François Marie Arouet de, 219

Wallace, Alfred Russel, 57
Walsh, Doris, 187
Ward, Francis Alan Burnett, 4, 8
Watches, spring-balance, 14. See also Clocks, Mechanical clocks
Waters, David W., 12, 14
Watson, Samuel, 146
Watt, James, 17, 146, 165, 202, 237
Watt, James, Jr., 192, 146–47, 153–54
Webster, John, 118–19
Webster, Noah, 132
Wedgwood, Josiah, 146, 202, 208, 210–11, 237
Weights and measures, 65–69, 234; light year, 20; barile, 65; bat, 65; batman, 65; Bible, 65–66; cubit, 66, 69, 72; ephah, 66; hin, 66; Magna Carta, 66; shekel, 66; British system, 66–74, 234; ell, 66–67; ancell, 67; avoirdupois, 67, 72–74; troy, 67, 73–74; yard, 67, 69, 71; balance, 68; beam-scale, 68; Sumerian, 68; digit, 69, 72, 77; foot, 69, 72, 78; hand, 69; nail, 69, 72; palm, 69, 72; fathom, 69–71; French metric system, 70–71, 74–77, 234; German, 70; Imperial Weights and Measures Act (1824), 70, 73–74; mile, 70–71; acre, 71; grain of barley, 71; rod, 71; ulna, 71; chain, 71; furlong, 71; morgen, 71; ton, 71; Roman, 71–72; nail or digit, 72; clove, 72; grain of wheat, 72; hundredweight, 72; inch, ounce, uncia, 69, 72, 79; pace, 72; penny, 72; stadium, 72; stone, 72; pound, 72–73; grain, 73; pennyweight, 73; American gallon, 74; ale gallon, 74; gallon, 74, 77; Imperial gallon, 74; Queen Anne's wine gallon, 74; wine gallon, 74; Winchester corn gallon, 74;

United States, 76. *See also* Metric system
Whitehurst, John, of Derby, 146
Whitney, Eli, 204, 224, 237
Whittemore, Meg, 185
Wholesale and retail distribution, 182–85, 238–39; franchising, 185–87, 190, 238–39
Wilberforce, Samuel, 59
William III, 121, 189
Williams, Tennessee, 182
Williamson, Joseph, 15
Wood, John, Sr., 228

Woolworth, Frank Winfield, 182–83
Wordsworth, William, 24, 108
Wright, Orville, 158
Wright, Wilbur, 158

Yeltsin, Boris, 178
Young, Edward, 56

Zamyatin, Yevgeny, 170
Zerubavel, Eviatar, 9, 34–37, 45–46
Zeus, 5, 43
Zoroastrians, 44
Zurvanites, 43–44

www.ingramcontent.com/pod-product-compliance
Lightning Source LLC
Chambersburg PA
CBHW030131240426
43672CB00005B/105